Design of Masonry Structures

DESIGN OF MASONRY STRUCTURES

Third edition of *Load Bearing Brickwork Design*

A. W. Hendry, B.Sc., Ph.D., D.Sc., F.I.C.E., F.I. Struct.E., F.R.S.E.
B. P. Sinha, B.Sc., Ph.D., F.I. Struct.E., F.I.C.E., C. Eng.
and
S. R. Davies, B.Sc., Ph.D., M.I.C.E., C.Eng.

Department of Civil Engineering
University of Edinburgh, UK

E & FN SPON
An Imprint of Chapman & Hall
London · Weinheim · New York · Tokyo · Melbourne · Madras

Published by E & FN Spon, an imprint of Chapman & Hall, 2–6 Boundary Row, London SE1 8HN, UK

Chapman & Hall, 2–6 Boundary Row, London SE1 8HN, UK

Chapman & Hall GmbH, Pappelallee 3, 69469 Weinheim, Germany

Chapman & Hall USA, 115 Fifth Avenue, New York, NY 10003, USA

Chapman & Hall Japan, ITP-Japan, Kyowa Building, 3F, 2-2-1 Hirakawacho, Chiyoda-ku, Tokyo 102, Japan

Chapman & Hall Australia, 102 Dodds Street, South Melbourne, Victoria 3205, Australia

Chapman & Hall India, R. Seshadri, 32 Second Main Road, CIT East, Madras 600 035, India

First edition 1997

First published as *Load Bearing Brickwork Design*

(First edition 1981. Second edition 1986)

Typeset in 10/12pt Palatino by Thomson Press (India)
Printed in Great Britain

ISBN 0 419 21560 3

A catalogue record for this book is available from the British Library

∞ Printed on acid-free paper, manufactured in accordance with ANSI/NISO Z39.48-1992 (Permanence of Paper) and ANSI/NISO Z39.48-1984 (Permanence of Paper)

Contents

Preface to the third edition

The first edition of this book was published in 1981 as *Load Bearing Brickwork Design*, and dealt with the design of unreinforced structural brickwork in accordance with BS 5628: Part 1. Following publication of Part 2 of this Code in 1985, the text was revised and extended to cover reinforced and prestressed brickwork, and the second edition published in 1987.

The coverage of the book has been further extended to include blockwork as well as brickwork, and a chapter dealing with movements in masonry structures has been added. Thus the title of this third edition has been changed to reflect this expanded coverage. The text has been updated to take account of amendments to Part 1 of the British Code, reissued in 1992, and to provide an introduction to the forthcoming Eurocode 6 Part 1-1, published in 1996 as ENV 1996-1-1. This document has been issued for voluntary use prior to the publication of EC6 as a European Standard. It includes a number of 'boxed' values, which are indicative: actual values to be used in the various countries are to be prescribed in a National Application Document accompanying the ENV.

Edinburgh, June 1996

Preface to the second edition

Part 2 of BS 5628 was published in 1985 and relates to reinforced and prestressed masonry which is now finding wider application in practice. Coverage of the second edition of this book has therefore been extended to include consideration of the principles and application of this form of construction.

Edinburgh, April 1987

Preface to the first edition

The structural use of brick masonry has to some extent been hampered by its long history as a craft based material and some years ago its disappearance as a structural material was being predicted. The fact that this has not happened is a result of the inherent advantages of brickwork and the design of brick masonry structures has shown steady development, based on the results of continuing research in many countries. Nevertheless, structural brickwork is not used as widely as it could be and one reason for this lies in the fact that design in this medium is not taught in many engineering schools alongside steel and concrete. To help to improve this situation, the authors have written this book especially for students in university and polytechnic courses in structural engineering and for young graduates preparing for professional examination in structural design.

The text attempts to explain the basic principles of brickwork design, the essential properties of the materials used, the design of various structural elements and the procedure in carrying out the design of a complete building. In practice, the basic data and methodology for structural design in a given material is contained in a code of practice and in illustrating design procedures it is necessary to relate these to a particular document of this kind. In the present case the standard referred to, and discussed in some detail, is the British BS 5628 Part 1, which was first published in 1978. This code is based on limit state principles which have been familiar to many designers through their application to reinforced concrete design but which are summarised in the text.

No attempt has been made in this introductory book to give extensive lists of references but a short list of material for further study is included which will permit the reader to follow up any particular topic in greater depth.

Preparation of this book has been based on a study of the work of a large number of research workers and practising engineers to whom the authors acknowledge their indebtedness. In particular, they wish to express their thanks to the following for permission to reproduce material from their publications, as identified in the text: British Standards Institution; Institution of Civil Engineers; the Building Research Establishment; Structural Clay Products Ltd.

Edinburgh, June 1981

A. W. Hendry
B. P. Sinha
S. R. Davies

Acknowledgements

Preparation of this book has been based on a study of the work of a large number of research workers and practising engineers, to whom the authors acknowledge their indebtedness. In particular, they wish to express thanks to the British Standards Institution, the Institution of Civil Engineers, the Building Research Establishment and Structural Clay Products Ltd for their permission to reproduce material from their publications, as identified in the text. They are also indebted to the Brick Development Association for permission to use the illustration of Cavern Walks, Liverpool, for the front cover.

Extracts from DD ENV 1996-1-1: 1995 are reproduced with the permission of BSI. Complete copies can be obtained by post from BSI Customer Services, 389 Chiswick High Road, London W4 4AL. Users should be aware that DD ENV 1996-1-1: 1995 is a prestandard; additional information may be available in the national foreword in due course.

1

Loadbearing masonry buildings

1.1 ADVANTAGES AND DEVELOPMENT OF LOADBEARING MASONRY

The basic advantage of masonry construction is that it is possible to use the same element to perform a variety of functions, which in a steel-framed building, for example, have to be provided for separately, with consequent complication in detailed construction. Thus masonry may, simultaneously, provide structure, subdivision of space, thermal and acoustic insulation as well as fire and weather protection. As a material, it is relatively cheap but durable and produces external wall finishes of very acceptable appearance. Masonry construction is flexible in terms of building layout and can be constructed without very large capital expenditure on the part of the builder.

In the first half of the present century brick construction for multi-storey buildings was very largely displaced by steel- and reinforced-concrete-framed structures, although these were very often clad in brick. One of the main reasons for this was that until around 1950 loadbearing walls were proportioned by purely empirical rules, which led to excessively thick walls that were wasteful of space and material and took a great deal of time to build. The situation changed in a number of countries after 1950 with the introduction of structural codes of practice which made it possible to calculate the necessary wall thickness and masonry strengths on a more rational basis. These codes of practice were based on research programmes and building experience, and, although initially limited in scope, provided a sufficient basis for the design of buildings of up to thirty storeys. A considerable amount of research and practical experience over the past 20 years has led to the improvement and refinement of the various structural codes. As a result, the structural design of masonry buildings is approaching a level similar to that applying to steel and concrete.

1.2 BASIC DESIGN CONSIDERATIONS

Loadbearing construction is most appropriately used for buildings in which the floor area is subdivided into a relatively large number of rooms of small to medium size and in which the floor plan is repeated on each storey throughout the height of the building. These considerations give ample opportunity for disposing loadbearing walls, which are continuous from foundation to roof level and, because of the moderate floor spans, are not called upon to carry unduly heavy concentrations of vertical load. The types of buildings which are compatible with these requirements include flats, hostels, hotels and other residential buildings.

The form and wall layout for a particular building will evolve from functional requirements and site conditions and will call for collaboration between engineer and architect. The arrangement chosen will not usually be critical from the structural point of view provided that a reasonable balance is allowed between walls oriented in the principal directions of the building so as to permit the development of adequate resistance to lateral forces in both of these directions. Very unsymmetrical arrangements should be avoided as these will give rise to torsional effects under lateral loading which will be difficult to calculate and which may produce undesirable stress distributions.

Stair wells, lift shafts and service ducts play an important part in deciding layout and are often of primary importance in providing lateral rigidity.

The great variety of possible wall arrangements in a masonry building makes it rather difficult to define distinct types of structure, but a rough classification might be made as follows:

- Cellular wall systems
- Simple or double cross-wall systems
- Complex arrangements.

A cellular arrangement is one in which both internal and external walls are loadbearing and in which these walls form a cellular pattern in plan. Figure 1.1(a) shows an example of such a wall layout.

The second category includes simple cross-wall structures in which the main bearing walls are at right angles to the longitudinal axis of the building. The floor slabs span between the main cross-walls, and longitudinal stability is achieved by means of corridor walls, as shown in Fig. 1.1(b). This type of structure is suitable for a hostel or hotel building having a large number of identical rooms. The outer walls may be clad in non-loadbearing masonry or with other materials.

It will be observed that there is a limit to the depth of building which can be constructed on the cross-wall principle if the rooms are to have

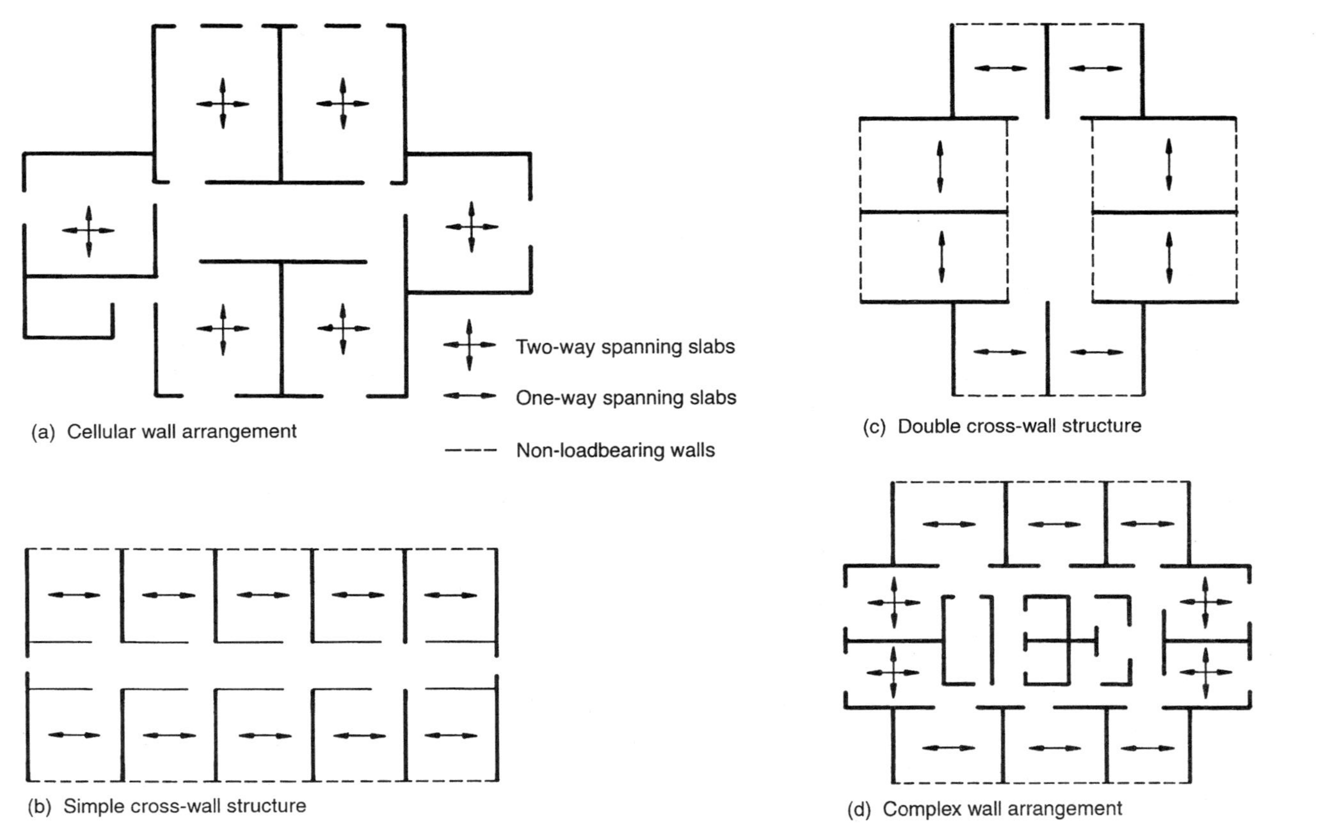

Fig. 1.1 Typical wall arrangements in masonry buildings.

effective day-lighting. If a deeper block with a service core is required, a somewhat more complex system of cross-walls set parallel to both major axes of the building may be used, as in Fig. 1.1(c).

All kinds of hybrids between cellular and cross-wall arrangements are possible, and these are included under the heading 'complex', a typical example being shown in Fig. 1.1(d).

Considerable attention has been devoted in recent years to the necessity for ensuring the 'robustness' of buildings. This has arisen from a number of building failures in which, although the individual members have been adequate in terms of resisting their normal service loads, the building as a whole has still suffered severe damage from abnormal loading, resulting for example from a gas explosion or from vehicle impact. It is impossible to quantify loads of this kind, and what is required is to construct buildings in such a way that an incident of this category does not result in catastrophic collapse, out of proportion to the initial forces. Meeting this requirement begins with the selection of wall layout since some arrangements are inherently more resistant to abnormal forces than others. This point is illustrated in Fig. 1.2: a building consisting only of floor slabs and cross-walls (Fig. 1.2(a)) is obviously unstable and liable to collapse under the influence of small lateral forces acting parallel to its longer axis. This particular weakness could be removed by incorporating a lift shaft or stair well to provide resistance in the weak direction, as in Fig. 1.2(b). However, the flank or gable walls are still vulnerable, for example to vehicle impact, and limited damage to this wall on the lowermost storey would result in the collapse of a large section of the building.

A building having a wall layout as in Fig. 1.2(c) on the other hand is clearly much more resistant to all kinds of disturbing forces, having a high degree of lateral stability, and is unlikely to suffer extensive damage from failure of any particular wall.

Robustness is not, however, purely a matter of wall layout. Thus a floor system consisting of unconnected precast planks will be much less resistant to damage than one which has cast-*in-situ* concrete floors with two-way reinforcement. Similarly, the detailing of elements and their connections is of great importance. For example, adequate bearing of beams and slabs on walls is essential in a gravity structure to prevent possible failure not only from local over-stressing but also from relative movement between walls and other elements. Such movement could result from foundation settlement, thermal or moisture movements. An extreme case occurs in seismic areas where positive tying together of walls and floors is essential.

The above discussion relates to multi-storey, loadbearing masonry buildings, but similar considerations apply to low-rise buildings where there is the same requirement for essentially robust construction.

Elevation

(a) Cross-walls without longitudinal walls: unstable

(b) Cross-walls with service shaft: normally stable but vulnerable to accidental damage

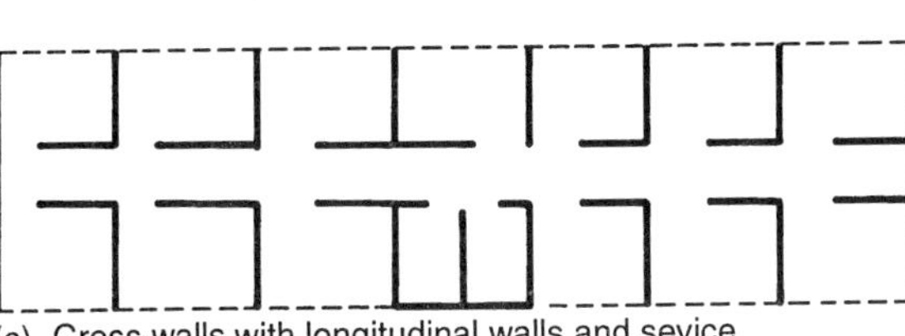

(c) Cross-walls with longitudinal walls and sevice shaft: robust construction

Structural walls

Non-structural walls

Fig. 1.2 Liability of a simple cross-wall structure to accidental damage.

1.3 STRUCTURAL SAFETY: LIMIT STATE DESIGN

The objective of ensuring a fundamentally stable or robust building, as discussed in section 1.2, is an aspect of structural safety. The measures adopted in pursuit of this objective are to a large extent qualitative and conceptual whereas the method of ensuring satisfactory structural performance in resisting service loads is dealt with in a more quantitative manner, essentially by trying to relate estimates of these loads with estimates of material strength and rigidity.

The basic aim of structural design is to ensure that a structure should fulfil its intended function throughout its lifetime without excessive deflection, cracking or collapse. The engineer is expected to meet this aim with due regard to economy and durability. It is recognized, however, that it is not possible to design structures which will meet these requirements in all conceivable circumstances, at least within the limits of financial feasibility. For example, it is not expected that normally designed structures will be capable of resisting conceivable but improbable accidents which would result in catastrophic damage, such as impact of a large aircraft. It is, on the other hand, accepted that there is uncertainty in the estimation of service loads on structures, that the strength of construction materials is variable, and that the means of relating loads to strength are at best approximations. It is possible that an unfavourable combination of these circumstances could result in structural failure; design procedures should, therefore, ensure that the probability of such a failure is acceptably small.

The question then arises as to what probability of failure is 'acceptably small'. Investigation of accident statistics suggests that, in the context of buildings, a one-in-a-million chance of failure leading to a fatality will be, if not explicitly acceptable to the public, at least such as to give rise to little concern. In recent years, therefore, structural design has aimed, indirectly, to provide levels of safety consistent with a probability of failure of this order.

Consideration of levels of safety in structural design is a recent development and has been applied through the concept of 'limit state' design. The definition of a limit state is that a structure becomes unfit for its intended purpose when it reaches that particular condition. A limit state may be one of complete failure (ultimate limit state) or it may define a condition of excessive deflection or cracking (serviceability limit state). The advantage of this approach is that it permits the definition of direct criteria for strength and serviceability taking into account the uncertainties of loading, strength and structural analysis as well as questions such as the consequences of failure.

The essential principles of limit state design may be summarized as follows. Considering the ultimate limit state of a particular structure, for failure to occur:

$$R^* - S^* \leqslant 0 \tag{1.1}$$

where $R^* = R_k / \gamma_m$ is the design strength of the structure, and $S^* = f(\gamma_f Q_k)$ the design loading effects. Here γ_m and γ_f are *partial safety factors*; R_k and Q_k are *characteristic values* of resistance and load actions, generally chosen such that 95% of samples representing R_k will exceed this value and 95% of the applied forces will be less than Q_k.

The probability of failure is then:

$$\mathrm{P}\,[R^* - S^* \leqslant 0] = p \tag{1.2}$$

If a value of p, say 10^{-6}, is prescribed it is possible to calculate values of the partial safety factors, γ_m and γ_f, in the limit state equation which would be consistent with this probability of failure. In order to do this, however, it is necessary to define the load effects and structural resistance in statistical terms, which in practice is rarely possible. The partial safety factors, therefore, cannot be calculated in a precise way and have to be determined on the basis of construction experience and laboratory testing against a background of statistical theory. The application of the limit state approach as exemplified by the British Code of Practice BS 5628 and Eurocode 6 (EC 6) is discussed in Chapter 4.

1.4 FOUNDATIONS

Building structures in loadbearing masonry are characteristically stiff in the vertical direction and have a limited tolerance for differential movement of foundations. Studies of existing buildings have suggested that the maximum relative deflection (i.e. the ratio of deflection to the length of the deflected part) in the walls of multi-storeyed loadbearing brickwork buildings should not exceed 0.0003 in sand or hard clay and 0.0004 in soft clay. These figures apply to walls whose length exceeds three times their height. It has also been suggested that the maximum average settlement of a brickwork building should not exceed 150 mm. These figures are, however, purely indicative, and a great deal depends on the rate of settlement as well as on the characteristics of the masonry. Settlement calculations by normal soil mechanics techniques will indicate whether these limits are likely to be exceeded. Where problems have arisen, the cause has usually been associated with particular types of clay soils which are subject to excessive shrinkage in periods of dry weather. In these soils the foundations should be at a depth of not less than 1 m in order to avoid moisture fluctuations.

High-rise masonry buildings are usually built on a reinforced concrete raft of about 600 mm thickness. The wall system stiffens the raft and helps to ensure uniform ground pressures, whilst the limitation on floor spans which applies to such structures has the effect of minimizing the amount of reinforcement required in the foundation slab. Under exceptionally good soil conditions it may be possible to use spread footings, whilst very unfavourable conditions may necessitate piling with ground beams.

1.5 REINFORCED AND PRESTRESSED MASONRY

The preceding paragraphs in this chapter have been concerned with the use of unreinforced masonry. As masonry has relatively low strength in tension, this imposes certain restrictions on its field of application. Concrete is, of course, also a brittle material but this limitation is overcome by the introduction of reinforcing steel or by prestressing. The corresponding use of these techniques in masonry construction is not new but, until recently, has not been widely adopted. This was partly due to the absence of a satisfactory code of practice, but such codes are now available so that more extensive use of reinforced and prestressed masonry may be expected in future.

By the adoption of reinforced or prestressed construction the scope of masonry can be considerably extended. An example is the use of prestressed masonry walls of cellular or fin construction for sports halls and similar buildings where the requirement is for walls some 10 m in height supporting a long span roof. Other examples include the use of easily constructed, reinforced masonry retaining walls and the reinforcement of laterally loaded walls to resist wind or seismic forces.

In appropriate cases, reinforced masonry will have the advantage over concrete construction of eliminating expensive shuttering and of producing exposed walls of attractive appearance without additional expense.

Reinforcement can be introduced in masonry elements in several ways. The most obvious is by placing bars in the bed joints or collar joints, but the diameter of bars which can be used in this way is limited. A second possibility is to form pockets in the masonry by suitable bonding patterns or by using specially shaped units. The steel is embedded in these pockets either in mortar or in small aggregate concrete (referred to in the USA as 'grout'). The third method, suitable for walls or beams, is to place the steel in the cavity formed by two leaves (or wythes) of brickwork which is subsequently filled with small aggregate concrete. This is known as grouted cavity construction. Elements built in this way can be used either to resist in-plane loading, as beams or shear walls, or as walls under lateral loading. In seismic situations it is possible to bond grouted cavity walls to floor slabs to give continuity to the structure. Finally, reinforcement can be accommodated in hollow block walls or piers, provided that the design of the blocks permits the formation of continuous ducts for the reinforcing bars.

Prestressed masonry elements are usually post-tensioned, the steel, in strand or bar form, being accommodated in ducts formed in the masonry. In some examples of cellular or diaphragm wall construction the prestressing steel has been placed in the cavity between the two masonry skins, suitably protected against corrosion. It is also possible to prestress circular tanks with circumferential wires protected by an outer skin of brickwork built after prestressing has been carried out.

2

Bricks, blocks and mortars

2.1 INTRODUCTION

Masonry is a well proven building material possessing excellent properties in terms of appearance, durability and cost in comparison with alternatives. However, the quality of the masonry in a building depends on the materials used, and hence all masonry materials must conform to certain minimum standards. The basic components of masonry are block, brick and mortar, the latter being in itself a composite of cement, lime and sand and sometimes of other constituents. The object of this chapter is to describe the properties of the various materials making up the masonry.

2.2 BRICKS AND BLOCKS

2.2.1 Classification

Brick is defined as a masonry unit with dimensions (mm) not exceeding 337.5 × 225 × 112.5 ($L \times w \times t$). Any unit with a dimension that exceeds any one of those specified above is termed a block. Blocks and bricks are made of fired clay, calcium silicate or concrete. These must conform to relevant national standards, for example in the United Kingdom to BS 3921 (clay units), BS 187 (calcium silicate) and BS 6073: Part 1 (concrete units). In these standards two classes of bricks are identified, namely common and facing; BS 3921 identifies a third category, engineering:

- *Common bricks* are suitable for general building work.
- *Facing bricks* are used for exterior and interior walls and available in a variety of textures and colours.
- *Engineering bricks* are dense and strong with defined limits of absorption and compressive strength as given in Table 2.2.

Bricks must be free from deep and extensive cracks, from damage to edges and corners and also from expansive particles of lime.

Bricks are also classified according to their resistance to frost and the maximum soluble salt content.

(a) Designation according to frost resistance

- *Frost resistant* (F): These bricks are durable in extreme conditions of exposure to water and freezing and thawing. These bricks can be used in all building situations.
- *Moderately frost resistant* (M): These bricks are durable in the normal condition of exposure except in a saturated condition and subjected to repeated freezing and thawing.
- *Not frost resistant* (O): These bricks are suitable for internal use. They are liable to be damaged by freezing and thawing unless protected by an impermeable cladding during construction and afterwards.

(b) Designation according to maximum soluble salt content

- *Low* (L): These clay bricks must conform to the limit prescribed by BS 3921 for maximum soluble salt content given in Table 2.1. All engineering and some facing or common bricks may come under this category.
- *Normal* (N): There is no special requirement or limit for soluble salt content.

2.2.2 Varieties

Bricks may be wire cut, with or without perforations, or pressed with single or double frogs or cellular. Perforated bricks contain holes; the cross-sectional area of any one hole should not exceed 10% and the volume of perforations 25% of the total volume of bricks. Cellular bricks will have cavities or frogs exceeding 20% of the gross volume of the brick. In bricks having frogs the total volume of depression should be

Table 2.1 Maximum salt content of low (L) brick (BS 3921)

Soluble radicals	*Maximum content as tested on 10 brick samples* (wt%)
Sulphate	0.50
Calcium	0.30
Magnesium	0.03
Potassium	0.03
Sodium	0.03

less than or equal to 20%. In the United Kingdom, calcium silicate or concrete bricks are also used, covered by BS 187 and BS 6073.

Bricks of shapes other than rectangular prisms are referred to as 'standard special' and covered by BS 4729.

Concrete blocks may be solid, cellular or hollow.

Different varieties of bricks and blocks are shown in Figs. 2.1 and 2.2.

2.2.3 Compressive strength

From the structural point of view, the compressive strength of the unit is the controlling factor. Bricks of various strengths are available to suit a wide range of architectural and engineering requirements. Table 2.2 gives a classification of bricks according to the compressive strength. For low-rise buildings, bricks of 5.2 N/mm^2 should be sufficient. For damp-proof courses, low-absorption engineering bricks are usually required. For reinforced and prestressed brickwork, it is highly unlikely that brick strength lower than 20 N/mm^2 will be used in the UK.

Calcium silicate bricks of various strengths are also available. Table 2.3 gives the class and strength of these bricks available.

Concrete bricks of minimum average strength of 21 to 50 N/mm^2 are available. Solid, cellular and hollow concrete blocks of various thicknesses and strengths are manufactured to suit the design requirements. Both the thickness and the compressive strength of concrete blocks are given in Table 2.4.

2.2.4 Absorption

Bricks contain pores; some may be 'through' pores, others are 'cul-de-sac' or even sealed and inaccessible. The 'through' pores allow air to escape in the 24 h absorption test (BS 3921) and permit free passage of water. However, others in a simple immersion test or vacuum test do not allow the passage of water, hence the requirement for a 5 h boiling or vacuum test. The absorption is the amount of water which is taken up to fill these pores in a brick by displacing the air. The saturation coefficient is the ratio of 24 h cold absorption to maximum absorption in vacuum or boiling. The absorption of clay bricks varies from 4.5 to 21% by weight and those of calcium silicate from 7 to 21% and concrete units 7 to 10% by weight. The saturation coefficient of bricks may range approximately from 0.2 to 0.88. Neither the absorption nor the saturation coefficient necessarily indicates the liability of bricks to decay by frost or chemical action. Likewise, absorption is not a mandatory requirement for concrete bricks or blocks as there is no relationship between absorption and durability.

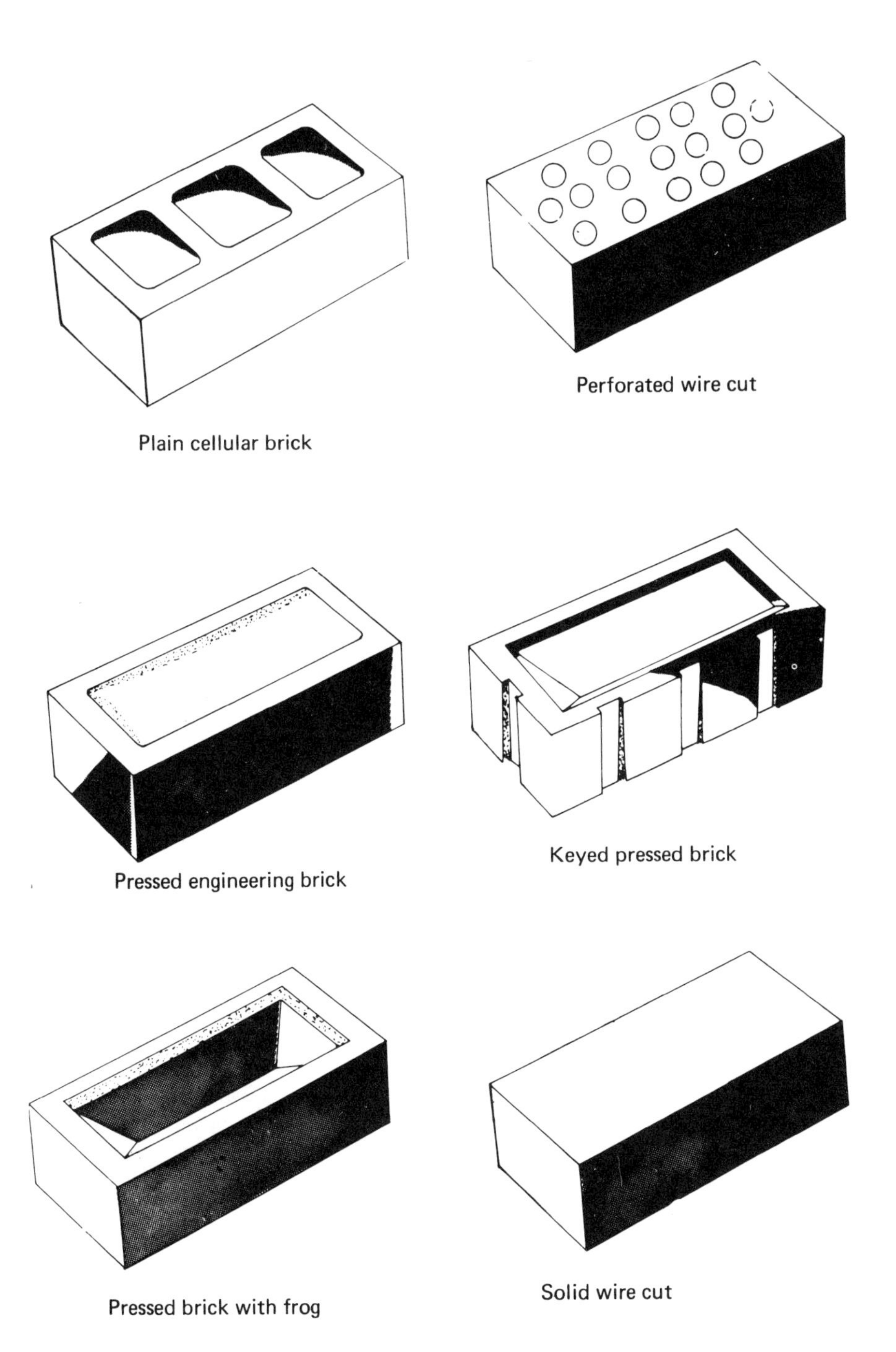

Fig. 2.1 Types of standard bricks.

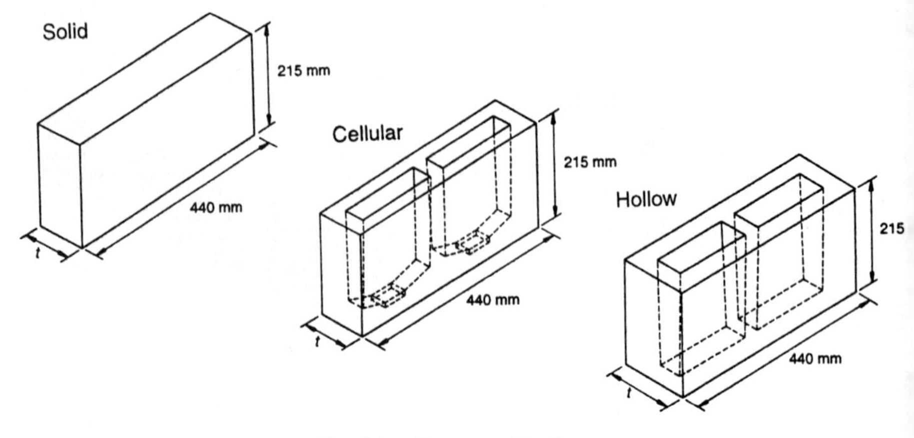

Fig. 2.2 Concrete blocks.

Table 2.2 Classification of clay bricks according to compressive strength and absorption

Designation	*Class*	*Average compressive strength not less than* (N/mm²)	*Average absorption (5 h boiling) not greater than* (% *by weight*)
Engineering	A	70	4.5
	B	50	7.0
Loadbearing brick		5–100	no specific requirement
Damp-proof course 1		5	4.5
Damp-proof course 2		5	7.0

Table 2.3 Compressive strength classes and requirements of calcium silicate bricks

Designation	*Class*	*Mean compressive strength of 10 bricks not less than* (N/mm²)	*Shrinkage not greater than* (%)	*Colour coding*
Loadbearing brick or Facing brick	7	48.5	0.040	green
	6	41.5		blue
	5	34.5		yellow
	4	27.5		red
	3	20.5		black
Facing brick or Common brick	2	14.0		–

Table 2.4 Compressive strength and thickness of concrete blocks

	Face size 440 × 215 mm	
Type	*Thickness* (mm)	*Minimum average compressive strength of unit* (N/mm^2)
Solid	75 100	
Solid or Cellular or Hollow	140 150 190 200 215	7.0–21.0

2.2.5 Frost resistance

The resistance of bricks to frost is very variable and depends on the degree of exposure to driving rain and temperature. Engineering bricks with high compressive strength and low absorption are expected to be frost resistant. However, some bricks of low strength and high absorption may be resistant to frost compared to low-absorption and high-strength brick.

Bricks can only be damaged provided 90% of the available pore space is filled with water about freezing temperature, since water expands one-tenth on freezing. Hence, low or high absorption of water by a brick does not signify that all the available pores will become filled with water. Calcium silicate bricks of 14 N/mm^2 or above are weather resistant.

In the United Kingdom, frost damage is not very common as brickwork is seldom sufficiently saturated by rain, except in unprotected cornices, parapets, free-standing and retaining walls. However, bricks and mortar must be carefully selected to avoid damage due to frost. Table 2.7 shows the minimum qualities of clay and calcium silicate bricks to be used for various positions in walls.

Precast concrete masonry units are frost resistant.

2.2.6 Dimensional changes

(a) Thermal movement

All building materials expand or contract with the rise and fall of temperature. The effect of this movement is dealt with in Chapter 13.

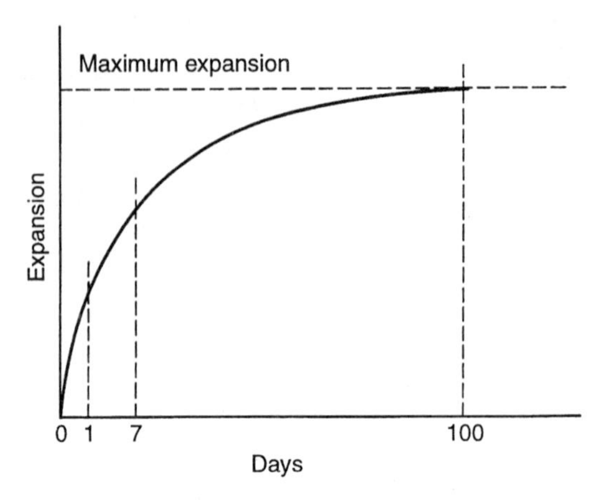

Fig. 2.3 Expansion of kiln-fresh bricks due to absorption of moisture from atmosphere.

(b) Moisture movement

One of the common causes of cracking and decay of building materials is moisture movement, which may be wholly or partly reversible or, in some circumstances, irreversible. The designer should be aware of the magnitude of this movement.

Clay bricks being taken from the kiln expand owing to absorption of water from the atmosphere. The magnitude of this expansion depends on the type of brick and its firing temperature and is wholly irreversible. A large part of this irreversible movement takes place within a few days, as shown in Fig. 2.3, and the rest takes place over a period of about six months. Because of this moisture movement, bricks coming fresh from the kiln should never be delivered straight to the site. Generally, the accepted time lag is a fortnight. Subsequent moisture movement is unlikely to exceed 0.02%.

In addition to this, bricks also undergo partly or wholly reversible expansion or contraction due to wetting or drying. This is not very significant except in the case of the calcium silicate bricks. Hence, the designer should incorporate 'expansion' joints in all walls of any considerable length as a precaution against cracking. Normally, movement joints in calcium silicate brickwork may be provided at intervals of 7.5 to 9.0 m depending upon the moisture content of bricks at the time of laying. In clay brickwork expansion joints at intervals of 12.2 to 18.3 m may be provided to accommodate thermal or other movements.

The drying shrinkage of concrete brick/blockwork should not exceed 0.06%. In concrete masonry, the movement joint should be provided at

Table 2.5 Moisture movement in different building materials

Materials	*Movement* (%)	
	Irreversible	*Reversible*
Clay bricks	0.10–0.20 (expansion)	negligible
Calcium silicate	0.001–0.05	0.001–0.05
Metal and glass	nil	nil
Dense concrete and mortar	0.02–0.12 (drying shrinkage)	0.01–0.055

6 m intervals as a general rule. However, the length of the panel without movement joint should not exceed twice the height.

Some indication of reversible or irreversible movement of various building materials is shown in Table 2.5.

The EC6 gives guidance for the design values of dimensional changes for unreinforced masonry, which are given in Chapter 4 (section 4.4).

2.2.7 Soluble salts

(a) Efflorescence

All clay bricks contain soluble salts to some extent. The salt can also find its way from mortar or soil or by contamination of brick by foreign agents. In a new building when the brickwork dries out owing to evaporation of water, the dissolved salts normally appear as a white deposit termed 'efflorescence' on the surface of bricks. Sometimes the colour may be yellow or pale green because of the presence of vanadium or chromium. The texture may vary from light and fluffy to hard and glassy. Efflorescence is caused by sulphates of sodium, potassium, magnesium and calcium; not all of these may be present in a particular case. Efflorescence can take place on drying out brickwork after construction or subsequently if it is allowed to become very wet. By and large, efflorescence does not normally result in decay, but in the United Kingdom, magnesium sulphate or sodium sulphate may cause disruption due to crystallization. Abnormal amounts of sodium sulphate, constituting more than 3% by weight of a brick, will cause disruption of its surface. Brick specimens showing efflorescence in the 'heavy' category are not considered to comply with BS 3921.

(b) Sulphate attack

Sulphates slowly react in the presence of water with tricalcium aluminate, which is one of the constituents of Portland cement and hydraulic lime. If water containing dissolved sulphate from clay bricks or aggregates reaches the mortar, this reaction takes place, causing mortar to crack and spall and thus resulting in the disintegration of the masonry. Sulphate attack is only possible if the masonry is exposed to very long and persistent wet conditions. Chimneys, parapets and earth-retaining walls which have not been properly protected from excessive dampness may be vulnerable to sulphate attack. In general, it is advisable to keep walls as dry as possible. In conditions of severe exposure to rain, bricks (L) or sulphate-resistant cement should be used. The resistance of mortar against sulphate attack can be increased by specifying a fairly rich mix, i.e. stronger than grade (iii) mortar (1:1:6) or replacing lime with a plasticizer. Calcium silicate and concrete units do not contain significant amounts of sulphate compared to clay bricks. However, concrete bricks of minimum 30 N/mm^2 strength should be used in $1:\frac{1}{2}:4\frac{1}{2}$ mortar for earth-retaining walls, cills and copings.

2.2.8 Fire resistance

Clay bricks are subjected to very much higher temperatures during firing than they are likely to be exposed to in a building fire. As a result, they possess excellent fire resistance properties. Calcium silicate bricks have similar fire resistance properties to clay bricks. Concrete bricks and blocks have 30 min to 6 h notional fire resistance depending on the thickness of the wall.

2.3 MORTAR

The second component in brickwork is mortar, which for loadbearing brickwork should be a cement:lime:sand mix in one of the designations shown in Table 2.6. For low-strength bricks a weaker mortar, 1:2:9 mix by volume, may be appropriate. For reinforced and prestressed brickwork, mortar weaker than grade (ii) ($1:\frac{1}{2}:4\frac{1}{2}$) is not recommended.

2.3.1 Function and requirement of mortar

In deciding the type of mortar the properties needing to be considered are:

- Development of early strength.
- Workability, i.e. ability to spread easily.
- Water retentivity, i.e. the ability of mortar to retain water against the suction of brick. (If water is not retained and is extracted quickly by a high-absorptive brick, there will be insufficient water left in the mortar joint for hydration of the cement, resulting in poor bond between brick and mortar.)

Table 2.6 Requirements for mortar (BS 5628)

Mortar designation	*Types of mortar (proportion by volume)*			*Mean compressive strength at 28 days* (N/mm²)	
	Cement:lime: sand	*Masonry cement: sand*	*Cement: sand with plasticizer*	*Preliminary (laboratory) test*	*Site tests*
(i)	1:0 to $\frac{1}{4}$:3	–	–	16.0	11.0
(ii)	1:$\frac{1}{2}$:4 to $4\frac{1}{2}$	1:$2\frac{1}{2}$ to $3\frac{1}{2}$	1:3 to 4	6.5	4.5
(iii)	1:1:5 to 6	1:4 to 5	1:5 to 6	3.6	2.5
(iv)	1:2:8 to 9	1:$5\frac{1}{2}$ to $6\frac{1}{2}$	1:7 to 8	1.5	1.0

↑ Increasing strength (upwards, from (iv) to (i))

↓ Increasing ability to accommodate movement, e.g. due to settlement, temperature and moisture changes (downwards, from (i) to (iv))

→ Increasing resistance to frost attack during construction

← Improvement in bond and consequent resistance to rain penetration

Direction of change in properties is shown by the arrows

- Proper development of bond with the brick.
- Resistance to cracking and rain penetration.
- Resistance to frost and chemical attack, e.g. by soluble sulphate.
- Immediate and long-term appearance.

2.3.2 Cement

The various types of cement used for mortar are as follows.

(a) Portland cement

Ordinary Portland cement and rapid-hardening cement should conform to a standard such as BS 12. Rapid-hardening cement may be used instead of ordinary Portland cement where higher early strength is required; otherwise its properties are similar. Sulphate-resistant cement should be used in situations where the brickwork is expected to remain wet for prolonged periods or where it is susceptible to sulphate attack, e.g. in brickwork in contact with sulphate-bearing soil.

(b) Masonry cement

This is a mixture of approximately 75% ordinary Portland cement, an inert mineral filler and an air-entraining agent. The mineral filler is used to reduce the cement content, and the air-entraining agent is added to improve the workability. Mortar made from masonry cement will have lower strength compared to a normal cement mortar of similar mix. The other properties of the mortar made from the masonry cement are intermediate between cement:lime:sand mortar and plasticized cement:sand mortar.

2.4 LIME: NON-HYDRAULIC OR SEMI-HYDRAULIC LIME

Lime is added to cement mortar to improve the workability, water retention and bonding properties. The water retentivity property of lime is particularly important in situations where dry bricks might remove a considerable amount of water from the mortar, thus leaving less than required for the hydration of the cement. Two types of lime are used, non-hydraulic or semi-hydraulic, as one of the constituents of mortar for brickwork. These limes are differentiated by the process whereby they harden and develop their strengths. Non-hydraulic lime initially stiffens because of loss of water by evaporation or suction by bricks, and eventually hardens because of slow carbonation, i.e. absorption of carbon dioxide from the air to change calcium hydroxide to calcium carbonate. Semi-hydraulic lime will harden in wet conditions as a result of the presence of small quantities of compounds of silica and alumina. It hardens owing to chemical reaction with water rather than atmospheric action. In the United Kingdom, the lime used for mortar must conform to BS 890.

2.5 SAND

The sand for mortar must be clean, sharp, and free from salt and organic contamination. Most natural sand contains a small quantity of silt or clay. A small quantity of silt improves the workability. Loam or clay is moisture-sensitive and in large quantities causes shrinkage of mortar. Marine and estuarine sand should not be used unless washed completely to remove magnesium and sodium chloride salts which are

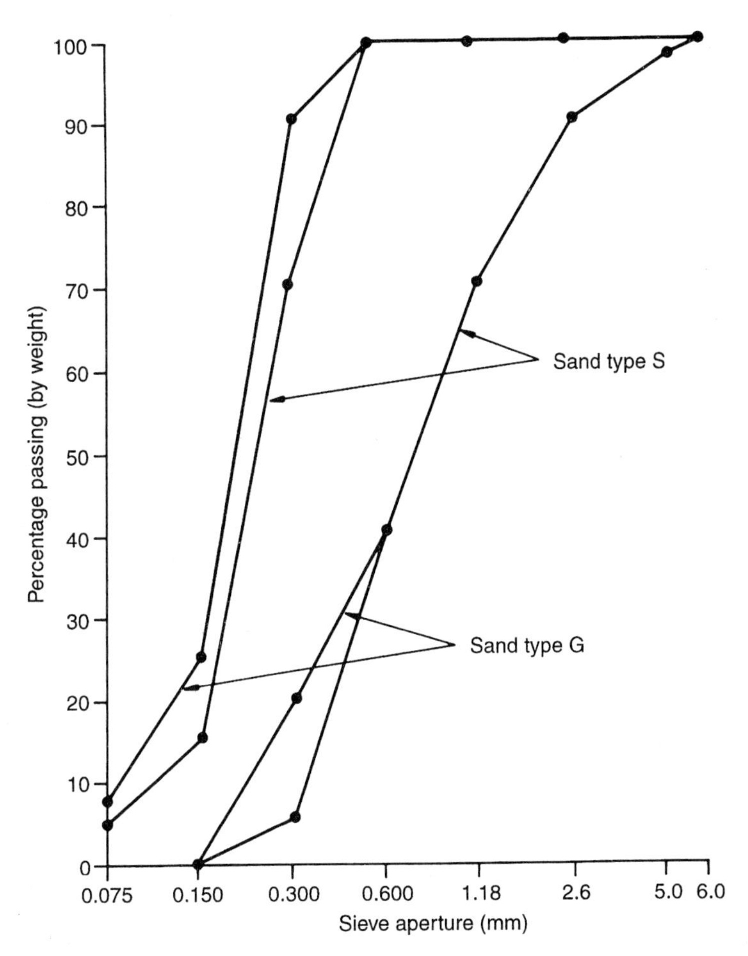

Fig. 2.4 Grading limits of mortar sand (BS 1200).

deliquescent and attract moisture. Specifications of sand used for mortar, such as BS 1200, prescribe grading limits for the particle size distribution. The limits given in BS 1200 are as shown in Fig. 2.4, which identifies two types of sand:sand type S and sand type G. Both types of sand will produce satisfactory mortars. However, the grading of sand type G, which falls between the lower limits of sand S and sand G, may require slightly more cement for a particular grade of mortar to satisfy the strength requirement envisaged in BS 5628 (refer to Table 2.6).

2.6 WATER

Mixing water for mortar should be clean and free from contaminants either dissolved or in suspension. Ordinary drinking water will be suitable.

2.7 PLASTICIZED PORTLAND CEMENT MORTAR

To reduce the cement content and to improve the workability, plasticizer, which entrains air, may be used. Plasticized mortars have poor water retention properties and develop poor bond with highly absorptive bricks. Excessive use of plasticizer will have a detrimental effect on strength, and hence manufacturers' instructions must be strictly followed. Plasticizer must comply with the requirements of BS 4887.

2.8 USE OF PIGMENTS

On occasion, coloured mortar is required for architectural reasons. Such pigments should be used strictly in accordance with the instructions of the manufacturer since excessive amounts of pigment will reduce the compressive strength of mortar and interface bond strength. The quantity of pigment should not be more than 10% of the weight of the cement. In the case of carbon black it should not be more than 3%.

2.9 FROST INHIBITORS

Calcium chloride or preparations based on calcium chloride should not be used, since they attract water and cause dampness in a wall, resulting in corrosion of wall ties and efflorescence.

2.10 PROPORTIONING AND STRENGTH

The constituents of mortar are mixed by volume. The proportions of material and strength are given in Table 2.6. For loadbearing brickwork the mortar must be gauged properly by the use of gauging boxes and preferably should be weigh-batched.

Recent research (Fig. 2.5) has shown that the water/cement ratio is the most important factor which affects the compressive strength of grades I, II and III mortars. In principle, therefore, it would be advisable for the structural engineer to specify the water/cement ratio for mortar to be used for structural brickwork; but, in practice, the water/cement ratio for a given mix will be determined by workability. There are various laboratory tests for measuring the consistency of mortar, and these have been related to workability. Thus in the United Kingdom, a dropping ball test is used in which an acrylic ball of 10 mm diameter is dropped on to the surface of a sample of mortar from a height of 250 mm. A ball penetration of 10 mm is associated with satisfactory workability. The test is, however, not used on site, and it is generally left to the bricklayer to adjust the water content to achieve optimum workability. This in fact achieves a reasonably consistent water/cement ratio which varies from one mix to another. The water/cement ratio for 10 mm ball penetration, representing satisfactory workability, has been indicated in Fig. 2.5 for the three usual mortar mixes.

It is important that the practice of adding water to partly set mortar to restore workability (known as 'knocking up' the mix) should be prevented.

2.11 CHOICE OF UNIT AND MORTAR

Table 2.7 shows the recommended minimum quality of clay or calcium silicate or concrete bricks/blocks and mortar grades which should be used in various situations from the point of view of durability.

2.12 WALL TIES

In the United Kingdom, external cavity walls are used for environmental reasons. The two skins of the wall are tied together to provide some degree of interaction. Wall ties for cavity walls should be galvanized mild steel or stainless steel and must comply to BS 1243. Three types of ties (Fig. 2.6) are used for cavity walls.

- Vertical twist type made from 20 mm wide, 3.2 to 4.83 mm thick metal strip
- 'Butterfly'– made from 3.15 mm wire
- Double-triangle type – made from 4.5 mm wire.

For loadbearing masonry vertical twist type ties should be used for maximum co-action. For a low-rise building, or a situation where large differential movement is expected or for reason of sound insulation, more flexible ties should be selected. In certain cases where large differential movements have to be accommodated, special ties or fixings have to be used (see Chapter 13). In specially unfavourable situations non-

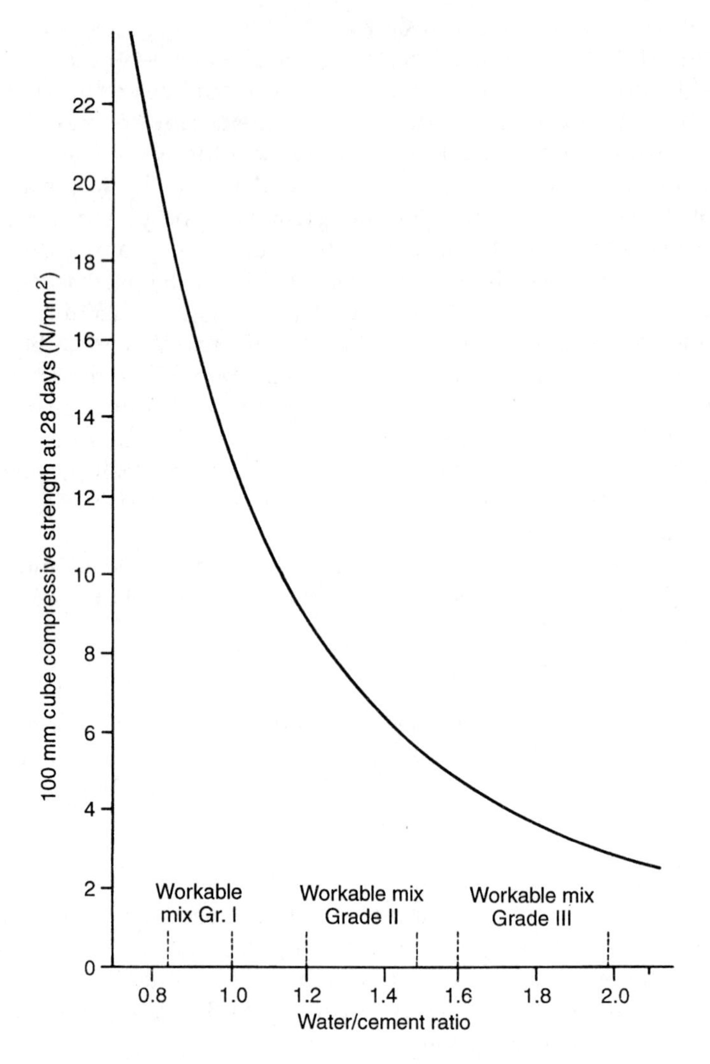

Fig. 2.5 Effect of water/cement ratio on the compressive strength of mortar of grades I, II and III.

ferrous or stainless-steel ties may be required. BS 5628 (Table 6) gives guidance for the selection and use of ties for normal situations.

2.13 CONCRETE INFILL AND GROUT

The mix proportion by volume for reinforced and prestressed masonry should be 1:0 to $\frac{1}{4}$:3:2 cement:lime:sand:10 mm maximum size aggregate.

Table 2.7 Durability of masonry in finished construction[a] (BS 5628)

(A) Work below or near external ground level

Masonry condition or situation	*Quality of masonry units and appropriate mortar designations*				*Remarks*
	Fired-clay units	*Calcium silicate*	*Concrete bricks*	*Concrete blocks*	
A1 Low risk of saturation with or without freezing	FL, FN, ML or MN in (i), (ii) or (iii)	Classes 3 to 7 in (iii) or (iv) (see remarks)	$\geqslant 15\,N/mm^2$ in (iii)	Either (a) of block density $\geqslant$ 1500 kg/m^3; or (b) made with dense aggregate complying with BS 882 or BS 1074; or (c) having a compressive strength $\geqslant 7\,N/mm^2$; or (d) most types of autoclaved aerated block (see remarks) in (iii)	Some types of autoclaved aerated concrete block may not be suitable. The manufacturer should be consulted. If sulphate ground conditions exist, the recommendations in **22.4** should be followed. Where designation (iv) mortar is used it is essential to ensure that all masonry units, mortar and masonry under construction are protected fully from saturation and freezing (see clause **30** and clause **35**).

Table 2.7 *(Contd)*

Masonry condition or situation	*Quality of masonry units and appropriate mortar designations*				*Remarks*
	Fired-clay units	*Calcium silicate*	*Concrete bricks*	*Concrete blocks*	
A2 High risk of saturation *without* freezing	FL, FN, ML or MN in (i) or (ii) (see remarks)	Classes 3 to 7 in (ii) or (iii)	$\geqslant 15\,N/mm^2$ in (ii) or (iii)	As for A1 in (ii) or (iii)	The masonry most vulnerable in A2 and A3 is located between 150 mm above, and 150 mm below, finished ground level. In this area masonry will become wet and may remain wet for long periods of time, particularly in winter. Where FN or MN fired-clay units are used in A2 or A3, sulphate-resisting cement should be used (see **22.4**).
A3 High risk of saturation *with* freezing	FL or FN in (i) or (ii)	Classes 3 to 7 in (ii)	$\geqslant 20\,N/mm^2$ in (ii) or (iii)	As for A1 in (ii)	
(B) Damp-proof courses					
B1 In buildings	Damp-proof course 1 as described in BS 3921, in (i)	Not suitable	Not suitable	Not suitable	Masonry DPCs can resist rising damp but will not resist water percolating downwards. If sulphate ground conditions exist, the recommendations in **22.4** should be followed.
B2 In external works	Damp-proof course 2 as described in BS 3921, in (i)	Not suitable	Not suitable	Not suitable	DPCs of fired-clay units are unlikely to be suitable for walls of other masonry units, as differential movement may occur (see **20.1**).

(C) Unrendered external walls (other than chimneys, cappings, copings, parapets, sills)					
C1 Low risk of situration	FL, FN, ML or MN in (i), (ii) or (iii) (see remarks)	Classes 2 to 7 in (iii) or (iv)	$\geqslant 7\,N/mm^2$ in (iii)	Any in (iii) or (iv) (see remarks)	Walls should be protected by roof overhang and other projecting features to minimize the risk of saturation. However, weathering details may not protect wall in conditions of very severe driving rain (see **21.3**). Certain architectural features, e.g brickwork below large glazed areas with flush sills, increase the risk of saturation (see **22.5**). Where designation(iv) mortar is used it is essential to ensure that all masonry units, mortar and masonry under construction are protected fully from saturation and freezing (see clause **30** and clause **35**). Where FN fired-clay units are used in designation (ii) mortar for C2, sulphate-resisting cement should be used (see **22.4**).
C2 High risk of saturation	FL or FN in (i) or (ii) (see remarks)	Classes 2 to 7 in (iii)	$\geqslant 15\,N/mm^2$ in (iii)	Any in (iii)	

Table 2.7 (*Contd*)

Masonry condition or situation	*Quality of masonry units and appropriate mortar designations*				*Remarks*
	Fired-clay units	*Calcium silicate*	*Concrete bricks*	*Concrete blocks*	
(D) Rendered external wall (other than chimneys, cappings, copings, parapets, sills)					
Rendered external walls (other than chimneys, cappings parapets, sills)	FN or MN in (i) or (ii) (see remarks) or FL or ML in (i) (ii) or (iii)	Classes 2 to 7 in (iii) or (iv) (see remarks)	$\geqslant 7\,N/mm^2$ in (iii)	Any in (ii) or (iv) (see remarks)	Rendered walls are usually suitable for most wind-driven rain conditions (see **21.3**). Where FN or MN fired-clay units are used, sulphate-resisting cement should be used in the mortar and in the base coat of the render (see **22.4**). Where designation (iv) mortar is used it is essential to ensure that all masonry units, mortar and masonry under construction are protected fully from saturation and freezing (see clauses **30** and **35**).
(E) Internal walls and inner leaves of cavity walls					
Internal walls and inner leaves of cavity walls	FL, FN, ML, MN, OL or ON in (i) (ii), (iii) or (iv) (see remarks)	Classes 2 to 7 in (iii) or (iv) (see remarks)	$\geqslant 7\,N/mm^2$ in (iv) (see remarks)	Any in (iii) or (iv) (see remarks)	Where designation (iv) mortar is used it is essential to ensure that all masonry units, mortar and masonry under construction are protected fully from saturation and freezing (see Clauses **30** and **35**).

(F) Unrendered parapets (other than cappings and copings)					
F1 Low risk of saturation, e.g. low parapets on some single. storey buildings	FL, FN, ML or MN in (i), (ii) or (iii)	Classes 3 to 7 in (iii)	$\geqslant 20\,N/mm^2$ in (iii)	Either (a) of block density $\geqslant 1500\,kg/m^3$ or (b) made with dense aggregate complying with BS 882 or BS 1047; or (c) having a compressive strength $\geqslant 7\,N/mm^2$; or (d) most types of autoclaved aerated block (see remarks) in (iii)	Most parapets are likely to be severely exposed irrespective of the climatic exposure of the building as a whole. Copings and DPCs should be provided wherever possible. Some types of autoclaved aerated concrete block may not be suitable. The manufacturer should be consulted. Where FN fired-clay units are used in F2, suphate-resisting cement should be used (see **22.4**).
F2 High risk of saturation, e.g. where a capping only is provided for the masonry	FL or FN in (i) or (ii) (see remarks)	Classes 3 to 7 in (iii)	$\geqslant 20\,N/mm^2$ in (iii)	As for F1 in (ii)	

Table 2.7 (*Contd*)

Masonry condition or situation	*Quality of masonry units and appropriate mortar designations*				*Remarks*
	Fired-clay units	*Calcium silicate*	*Concrete bricks*	*Concrete blocks*	
(G) Rendered parapets (other than cappings and copings)					
Rendered parapets (other than cappings and copings)	FN or MN in (i) or (ii) (see remarks) *or* FL or ML (i), (ii) or (iii)	Classes 3 to 7 in (iii)	$\geqslant 7\,N/mm^2$ in (iii)	Any in (iii)	Single-leaf walls should be rendered only on one face. All parapets should be provided with a coping. Where FN or MN fired-clay units are used, sulphate-resisting cement should be used in the mortar *and* in the base coat of the render (see **22.4**).
(H) Chimneys					
H1 Unrendered with low risk of saturation	FL,FN,ML or MN in (i), (ii) or (iii)	Classes 3 to 7 in (iii)	$\geqslant 10\,N/mm^2$	Any in (iii)	Chimney stacks are normally the most exposed masonry on any building. Due to the possibility of sulphate attack from flue gases the use of sulphate-resisting cement in the mortar *and* in any render is strongly recommended (see **22.4**).
H2 Unrendered with high risk of saturation	FL or FN in (i) or (ii)	Classes 3 to 7 in (iii)	$\geqslant 15\,N/mm^2$ in (iii)	Either (a) of block density $\geqslant 1500\,kg/m^3$; or	

				(b) made with dense aggregate complying with BS 882 or BS 1047; or (c) having a compressive strength $\geqslant 7\ N/mm^2$; or (d) most types of autoclaved aerated block (see remarks) in (ii)	Brickwork and tile cappings cannot be relied upon to keep out moisture indefinitely. The use of a coping is preferable. Some types of autoclaved aerated concrete block may not be suitable for use in H2. The manufacturer should be consulted.
H3 Rendered	FL or ML in (i), (ii) or (iii) *or* FN or MN in (i) *or* (iii)	Classes 3 to 7 in (iii)	$\geqslant 7\ N/mm^2$ in (iii)	Any in (iii)	
(I) Cappings, copings and sills					
Cappings, copings and sills	FL or FN in (i)	Classes 4 to 7 in (ii)	$\geqslant 30\ N/mm^2$ in (ii)	Either (a) of block density $\geqslant 1500\ kg/m^3$; or (b) made with dense aggregate complying with BS 882 or BS 1047; or (c) having a compressive strength $\geqslant 7\ N/mm^2$; or (d) most autoclaved aerated blocks (see remarks) in (ii)	Some autoclaved aerated concrete block may be unsuitable for use I. The manufacturere shoule be consulted. Where cappings or copings are used for chimney terminals, the use of sulphate-resisting cement is strongly recommended (see **22.4**). DPCs for cappings, copings and sills should be bedded in the same mortar as the masonry units.

Table 2.7 (*Contd*)

(J) Free-standing boundary and screen walls (other than cappings and copings)

Masonry condition or situation	*Quality of masonry units and appropriate mortar designations*				*Remarks*
	Fired-clay units	*Calcium silicate*	*Concrete bricks*	*Concrete blocks*	
J1 With coping	FN or MN in (i) or (ii) *or* FL on ML in (i), (ii) or (iii)	Classes 3 to 7	$\geqslant 15\,N/mm^2$	Any in (iii)	Masonry in free-standing walls is likely to be severely exposed, irrespective of climatic conditions. Such walls should be protected by a coping wherever possible and DPCs should be provided under the copings *and* at the base of the wall (see clause **21**). Where FN or MN fired-clay units are used for J1 in conditions of severe driving rain (see clause **21**), the use of sulphate-resisting cement is strongly recommended (see **22.4**). Where designation (iii) mortar is in (ii)used for J2 the use of sulphate-resisting cement is strongly recommended (see **22.4**). Some types of autoclaved aerated concrete block may also be unsuitable. The manufacturer should be consulted.
J2 With capping	FL or FN in (i) or (ii) (see remarks)	Classes 3 to 7 in (iii)	$\geqslant 20\,N/mm^2$ in (iii)	Either (a) of block density $1500\,kg/m^3$; or (b) made with dense aggregate complying with BS 882 or BS 1047; or (c) having a compressive strength $\geqslant 7\,N/mm^2$ (see remarks) or (d) most types of autoclaved aerated block (see remarks) in (ii)	

(K) Earth-retaining walls (other than cappings and copings)					
K1 With water-proofed retaining face and coping	FL, FN, ML or MN in (i) or (ii)	Classes 3 to 7 in (ii) or (iii)	$\geqslant 15\,N/mm^2$ in (ii)	Either (a) of block density $\geqslant 1500\,kg/mm^3$; or (b) made with dense aggregate complying with BS 882 or BS 1047; or (c) having a compressive strength $\geqslant 7\,N/mm^2$; or (d) most types of autoclaved aerated block (see remarks) in (ii)	Because of possible contamination from the ground and saturation by ground waters, in addition to subjection to severe climatic exposure, masonry in retaining walls is particularly prone to frost and sulphate attack. Careful choice of materials in relation to the methods for exclusion of water recommended in clause **21** is essential. It is strongly recommended that such walls be backfilled with free-draining material. The provision of an effective coping with a DPC (see clause **21**) and waterproofing of the retaining face of the wall (see **22.1.2**) is desirable. Where FN or MN fired-clay units are used, the use of sulphate-resisting cement may be necessary (see **22.4**). Some types of autoclaved aerated concrete block are not suitable for use in K1. The manufacturer should be consulted.

Table 2.7 *(Contd)*

Masonry condition or situation	*Quality of masonry units and appropriate mortar designations*				*Remarks*
	Fired-clay units	*Calcium silicate*	*Concrete bricks*	*Concrete blocks*	
K2 With coping or capping but no waterproofing on retaining face	FL or FN in (i)	Classes 4 to 7 in (ii)	$\geqslant 30\ N/mm^2$ in (i) or (ii)	As for K1 but in (i) or (ii) (see remarks)	Most concrete blocks are not suitable for use in K2. The manufacturer should be consulted.
(L) Drainage and sewerage, e.g. inspection chambers, manholes					
L1 Surface water	Engineering bricks, FL, FN, ML or MN (see remarks) in (i)	Classes 3 to 7 in (ii) and (iii)	$\geqslant 20\ N/mm^2$ in (iii)	Either (a) of block density $\geqslant 1500\ kg/m^3$; or (b) made with dense aggregate complying with BS 882 or BS 1047; or (c) having a compressive strength $\geqslant 7\ N/mm^2$; or (d) most types of autoclaved aerated block (see remarks) in (ii)	Where FN fired-clay units are used, sulphate-resisting cement should be used. If sulphate ground conditions exist the recommendations in **22.4** should be followed. Some types of autoclaved aerated block are not suitable for use in L1. The manufacturer should be consulted. Some types of calcium silicate brick are not suitable for use in L2 or L3. The manufacturer should be consulted.

L2 Foul drainage (continuous contact with masonry)	Engineering bricks, FL, FN, ML or MN in (i)	Class 7 in (ii) (see remarks)	$\geqslant 40$ N/mm^2 with cement content $\geqslant 350$ kg/m^3 in (i) or (ii)	Not suitable
L3 Foul drainage (occasional contact with masonry	Engineering bricks, FL, FN, ML or MN in (i)	Classes 3 to 7 in (ii) and (iii) (see remarks)	$\geqslant 40$ N/mm^2 with cement content $\geqslant 350$ kg/m^3 in (i) or (ii)	Not suitable

[a] The numbers (i) to (iv) refer to mortar designations in Table 2.6. The classes 2 to 7 are described in Table 2.3. The clause numbers in **bold** in the 'Remarks' refer to BS 5628.

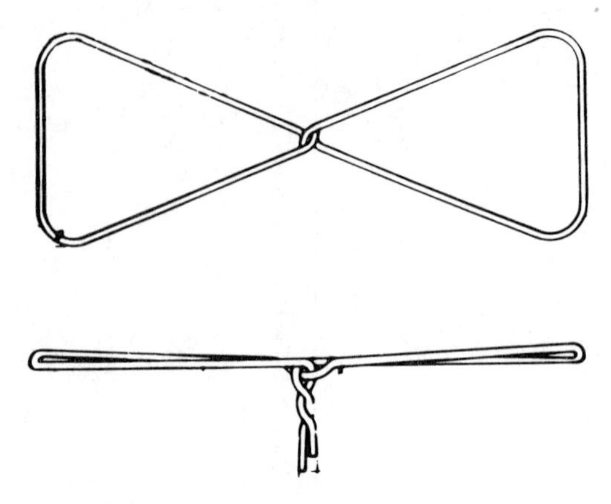

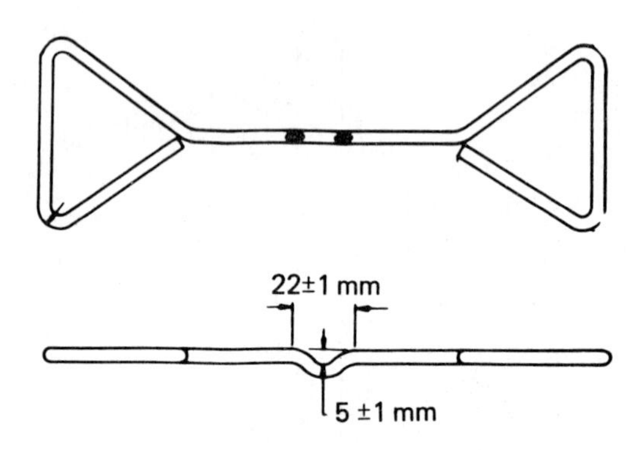

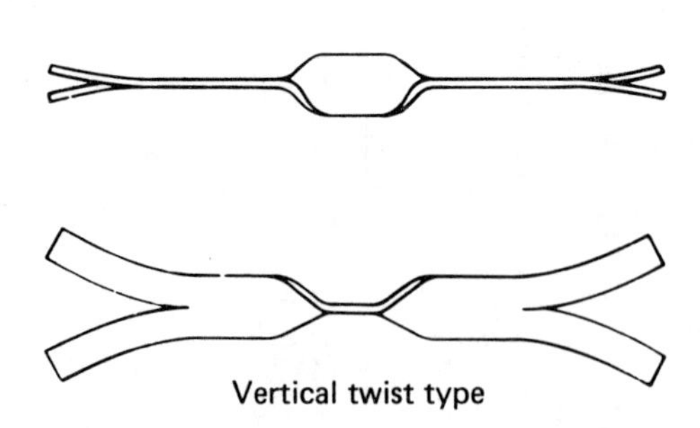

Fig. 2.6 Metal wall ties suitable for cavity walls.

Table 2.8 Chloride content of mixes

Type or use of concrete or mortar	*Maximum total chloride content by mass of cement* (mass %)
Prestressed concrete, heat-cured concrete containing embedded metal	0.1
Concrete or mortar made with cement complying with BS 4027	0.2
Concrete or mortar containing embedded metal and made with cement complying with BS 12 or BS 146	0.4

Table 2.9 Characteristic tensile strength of reinforcing steel

Designation	*Nominal size*	*Characteristic tensile strength, f_y* (N/mm^2)
Hot-rolled plain steel bars complying with BS 4449	all	250
Hot-rolled deformed high-yield steel bars complying with BS 4449	all	460
Cold-worked steel bars complying with BS 4461	all	460
Hard-drawn steel wire complying with BS 4482 and steel fabric complying with BS 4483	up to and including 12	485
Stainless steel complying with BS 970: Part 1 grades 316 S31 or 316 S33	all	460

The maximum size of the aggregate can be increased depending on the size and configuration of the void to be filled with concrete. In some cases it would be possible to use concrete design mix as specified in BS 5328 for reinforced and prestressed masonry. In reinforced and prestressed masonry, the bricks or blocks coming in contact with concrete will absorb water from the mix depending on its water retentivity property, and hence maximum free water/cement ratio used in BS 8110 may not be applicable. In order to compensate for this and for free flowing of the mix to fill the space and the void, a slump of 75 mm and 175 mm for concrete mix has been recommended in BS 5628: Part 2.

In prestressed sections where tendons are placed in narrow ducts, a neat cement or sand:cement grout having minimum compressive strength of 17 N/mm^2 at 7 days may be used.

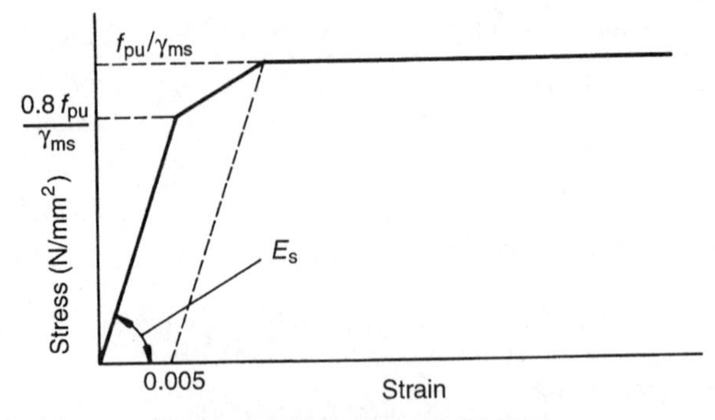

E_s 205 kN/mm^2 for cold drawn wire complying with BS 5896
195 kN/mm^2 for strand complying with BS 5896
165 kN/mm^2 for rolled and stretched bars complying with BS 4486
206 kN/mm^2 for rolled and as rolled stretched and tempered bars complying with BS 4486

f_{pu} is the characteristic tensile strength of prestressing tendons
γ_{ms} is the partial safety factor for strength of steel
E_s is the modulus of elasticity of steel

Fig. 2.7 Typical short-term design stress–strain curve for normal and low-relaxation tendons.

The mix must conform to the limit prescribed by BS 5628: Part 2 for maximum total chloride content as in Table 2.8.

2.14 REINFORCING AND PRESTRESSING STEEL

2.14.1 Reinforcing steel

Hot-rolled or cold-worked steel bars and fabric conforming to the relevant British Standard can be used as reinforcement. The characteristic strengths of reinforcement are given in Table 2.9.

In situations where there is risk of contamination by chloride, solid stainless steel or low-carbon steel coated with at least 1 mm of austenitic stainless steel may be used.

2.14.2 Prestressing steel

Wire, strands and bars complying to BS 4486 or BS 5896 can be used for prestressing. Seventy per cent of the characteristic breaking load is allowed as jacking force for prestressed masonry which is less than the 75% normally allowed in prestressed concrete. If proper precautions are taken, there is no reason why the initial jacking force cannot be taken to 75–80% of the breaking load. This has been successfully demonstrated in a series of prestressed brick test beams at Edinburgh University.

The short-term design stress–strain curve for prestressing steel is shown in Fig. 2.7.

3

Masonry properties

3.1 GENERAL

Structural design in masonry requires a clear understanding of the behaviour of the composite unit–mortar material under various stress conditions. Primarily, masonry walls are vertical loadbearing elements in which resistance to compressive stress is the predominating factor in design. However, walls are frequently required to resist horizontal shear forces or lateral pressure from wind and therefore the strength of masonry in shear and in tension must also be considered.

Current values for the design strength of masonry have been derived on an empirical basis from tests on piers, walls and small specimens. Whilst this has resulted in safe designs, it gives very little insight into the behaviour of the material under stress so that more detailed discussion on masonry strength is required.

3.2 COMPRESSIVE STRENGTH

3.2.1 Factors affecting compressive strength

The factors set out in Table 3.1 are of importance in determining the compressive strength of masonry.

Table 3.1 Factors affecting masonry strength

Unit characteristics	*Mortar characteristics*	*Masonry*
Strength	Strength:	Bond
Type and geometry:	mix	Direction of stressing
solid	water/cement ratio	Local stress raisers
perforated	water retentivity	
hollow	Deformation characteristics	
height/thickness ratio	relative to unit	
absorption characteristics		

3.2.2 Unit/mortar/masonry strength relationship

A number of important points have been derived from compression tests on masonry and associated standard tests on materials. These include, first, that masonry loaded in uniform compression will fail either by the development of tension cracks parallel to the axis of loading or by a kind of shear failure along certain lines of weakness, the mode of failure depending on whether the mortar is weak or strong relative to the units. Secondly, it is observed that the strength of masonry in compression is smaller than the nominal compressive strength of the units as given by a standard compressive test. On the other hand, the masonry strength may greatly exceed the cube crushing strength of the mortar used in it. Finally, it has been shown that the compressive strength of masonry varies roughly as the square root of the nominal unit crushing strength and as the third or fourth root of the mortar cube strength.

From these observations it may be inferred that:

1. The secondary tensile stresses which cause the splitting type of failure result from the restrained deformation of the mortar in the bed joints of the masonry.
2. The apparent crushing strength of the unit in a standard test is not a direct measure of the strength of the unit in the masonry, since the mode of failure is different in the two situations.
3. Mortar withstands higher compressive stresses in a brickwork bed joint because of the lateral restraint on its deformation from the unit.

Various theories for the compressive strength of masonry have been proposed based on equation of the lateral strains in the unit and mortar at their interface and an assumed limiting tensile strain in the unit. Other theories have been based on measurement of biaxial and triaxial strength tests on materials. But in both approaches the difficulties of determining the necessary materials properties have precluded their practical use, and for design purposes reliance continues to be placed on empirical relationships between unit, mortar and masonry strengths. Such relationships are illustrated in Fig. 3.1 and are incorporated in codes of practice, as set out in Chapter 4 for BS 5628 and Eurocode 6.

3.2.3 Some effects of unit characteristics

The apparent strength of a unit of given material increases with decrease in height because of the restraining effect of the testing machine platens on the lateral deformation of the unit. Also, in masonry the units have to resist the tensile forces resulting from restraint of the lateral strains in the mortar. Thus for given materials and joint thickness, the greater the height of the unit the greater the resistance to these forces and the

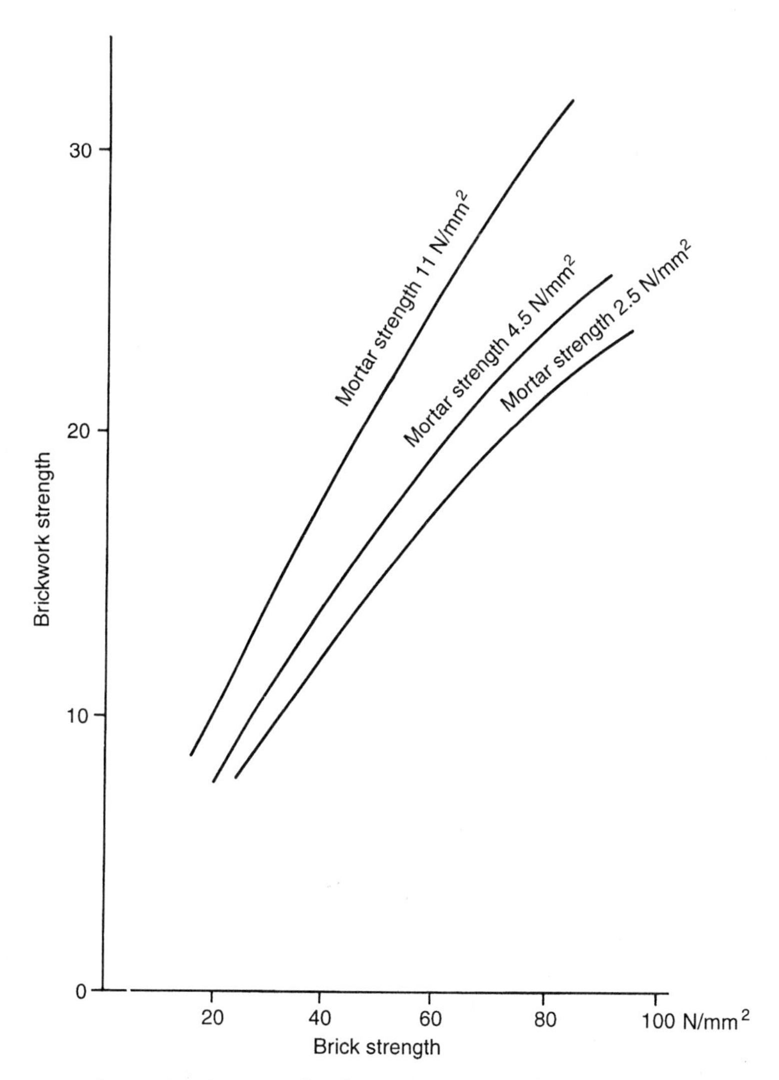

Fig. 3.1 Relationship between brick crushing strength and brickwork strength for various mortar strengths. Based on test results.

greater the compressive strength of the masonry. A corollary of this proposition is that, for a given unit height, increasing the thickness of the mortar joint will decrease the strength of the masonry. This effect is significant for brickwork, as shown in Fig. 3.2, but unimportant in blockwork where the ratio of joint thickness to unit height is small.

It follows from this discussion that the shape of a unit influences the strength of masonry built from it, and if units are laid on edge or on end

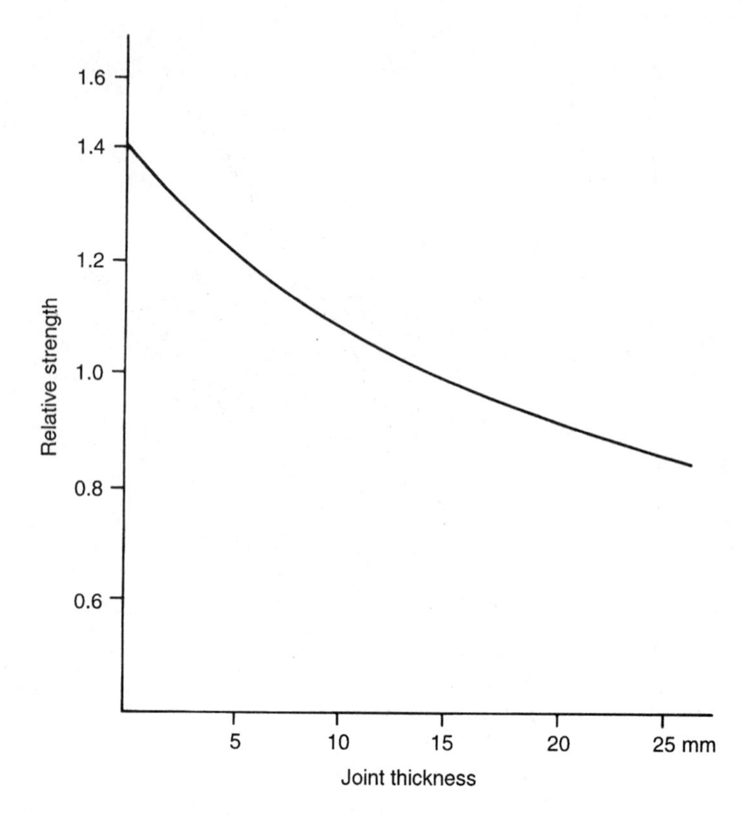

Fig. 3.2 Effect of joint thickness on brickwork strength.

the resulting masonry strength will be different from that of masonry in which the units are laid on their normal bed faces. The masonry strength will also depend on the type of unit: a highly perforated unit is likely to be relatively weak when compressed in a direction parallel to its length and thus result in a correspondingly lower masonry strength. This is illustrated in Table 3.2 which gives some results for brickwork built with various types of bricks. From this table it can be seen that, although there is a substantial reduction in brickwork strength when built and stressed in directions other than normal, this is not proportional to the brick strength when the latter is compressed in the corresponding direction. No general rule can be given relating brickwork to brick strength when compressed with the units laid on edge or on end.

Special considerations apply to masonry built with hollow blocks in which the cores may be unfilled or filled with concrete. In the former case the mortar joint may cover the whole of the bed face of the block (full-bedded) or only the outer shells (shell-bedded).

The strength of full-bedded blocks is taken to be that of the maximum test load divided by the gross area of the unit and the masonry strength

Table 3.2 Compressive strength of bricks and prisms compressed in different directions[a]

(a) Brick strength (N/mm^2)

Brick type	*Tested*		
	On bed	*On edge*	*On end*
14 hole	74.3 (100)	26.2 (35)	10.4 (14)
10 hole	70.2 (100)	29.5 (42)	21.7 (31)
3 hole	82.0 (100)	53.2 (65)	40.2 (49)
5 slots	64.1 (100)	51.8 (81)	13.8 (22)

(b) Prism strength (N/mm^2)

Brick on end	*Laid*		
	On bed	*On edge*	*On end*
14 hole	28.9 (100)	8.5 (29)	14.6 (51)
10 hole	22.0 (100)	15.0 (66)	20.0 (91)
3 hole	37.6 (100)	30.5 (78)	21.8 (56)
5 slots	34.1 (100)	29.0 (85)	13.9 (41)

[a] Figures in brackets indicate relative strengths. Mortar $1:\frac{1}{4}:3$.

is calculated as if the unit was solid. The strength of shell-bedded masonry should be calculated on the basis of the mortared area of the units.

Conventionally, the compressive strength of hollow block masonry built with the cores filled with concrete is taken to be the sum of the strengths of the hollow block and the concreted core tested separately. However, even when the materials are of approximately the same nominal strength, this rule is not always reliable as there can be a difference in the lateral strains of the block and fill materials at the ultimate load, resulting in a tendency for the fill to split the block. Various formulae have been devised to calculate the strength of filled block masonry, as for example the following which has been suggested by Khalaf (1991) to give the prism strength (f'_m) of this type of masonry:

$$f'_m = 0.3 f_b + 0.1 f_{mr} + 0.25 f_c \tag{3.1}$$

where f_b is the unit material compressive strength, f_{mr} is the mortar compressive strength and f_c is the core infill compressive strength.

3.3 STRENGTH OF MASONRY IN COMBINED COMPRESSION AND SHEAR

The strength of masonry in combined shear and compression is of importance in relation to the resistance of buildings to lateral forces. Many tests on masonry panels subjected to this type of loading have been carried out with a view to establishing limiting stresses for use in design. Typical results are shown in Fig. 3.3. It is found that there is a Coulomb type of relationship between shear strength and precompression, i.e. there is an initial shear resistance dependent on adhesion between the units and mortar augmented by a frictional component proportional to the precompression. This may be expressed by the formula:

$$\tau = \tau_0 + \mu \sigma_c \tag{3.2}$$

where τ_0 is the shear strength at zero precompression, μ is an apparent coefficient of friction and σ_c is the vertical compressive stress.

This relationship holds up to a certain limiting value of the vertical compression, beyond which the joint failure represented by the Coulomb equation is replaced by cracking through the units. For clay bricks this limit is about 2.0 N/mm^2. The shear strength depends on the mortar strength and for units with a compressive strength between 20 and 50 N/mm^2 set in strong (1:$\frac{1}{4}$:3) mortar the value of τ_0 will be approximately 0.3 N/mm^2 and 0.2 N/mm^2 for medium strength (1:1:6) mortar. The average value of μ is 0.4–0.6.

The shear stresses quoted above are average values for walls having a height-to-length ratio of 1.0 or more and the strength of a wall is calculated on the plan area of the wall in the plane of the shear force.

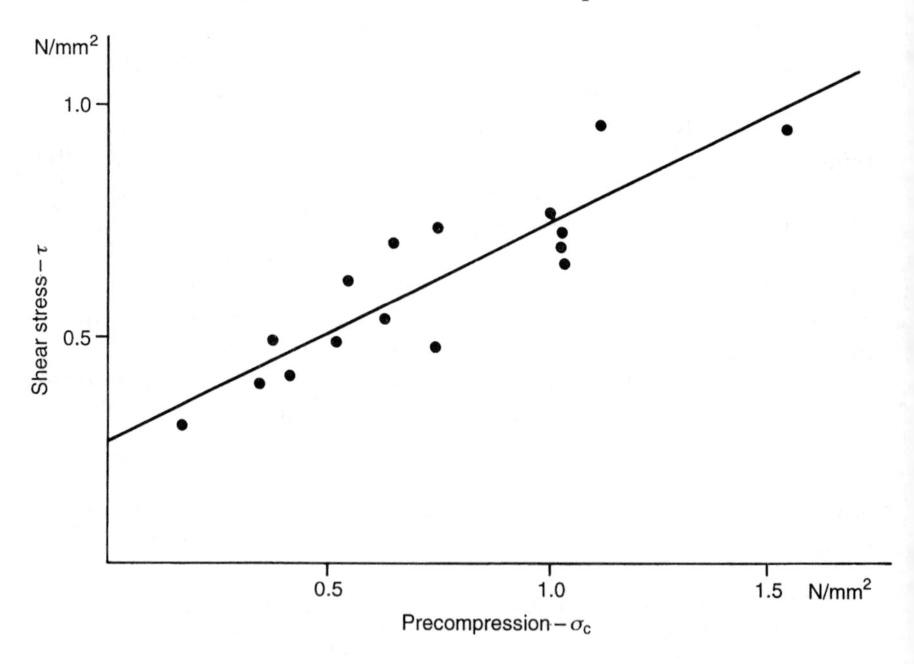

Fig. 3.3 Typical relationship between shear strength of brickwork and vertical precompression from test results.

That is to say, if a wall has returns at right angles to the direction of the shear force, the area of the returns is neglected in calculating the shear resistance of the wall.

3.4 THE TENSILE STRENGTH OF MASONRY

3.4.1 Direct tensile strength

Direct tensile stresses can arise in masonry as a result of in-plane loading effects. These may be caused by wind, by eccentric gravity loads, by thermal or moisture movements or by foundation movement. The tensile resistance of masonry, particularly across bed joints, is low and variable and therefore is not generally relied upon in structural design. Nevertheless, it is essential that there should be some adhesion between units and mortar, and it is necessary to be aware of those conditions which are conducive to the development of mortar bond on which tensile resistance depends.

The mechanism of unit–mortar adhesion is not fully understood but is known to be a physical–chemical process in which the pore structure of both materials is critical. It is known that the grading of the mortar sand is important and that very fine sands are unfavourable to adhesion. In the case of clay brickwork the moisture content of the brick at the time of laying is also important: both very dry and fully saturated bricks lead to low bond strength. This is illustrated in Fig. 3.4, which shows the results of bond tensile tests at brick moisture contents from oven-dry to fully saturated. This diagram also indicates the great variability of tensile bond strength and suggests that this is likely to be greatest at a moisture content of about three-quarters of full saturation, at least for the bricks used in these tests.

Direct tensile strength of brickwork is typically about $0.4\,N/mm^2$, but the variability of this figure has to be kept in mind, and it should only be used in design with great caution.

3.4.2 Flexural tensile strength

Masonry panels used essentially as cladding for buildings have to withstand lateral wind pressure and suction. Some stability is derived from the self-weight of a wall, but generally this is insufficient to provide the necessary resistance to wind forces, and therefore reliance has to be placed on the flexural tensile strength of the masonry.

The same factors as influence direct tensile bond, discussed in the preceding section, apply to the development of flexural tensile strength.

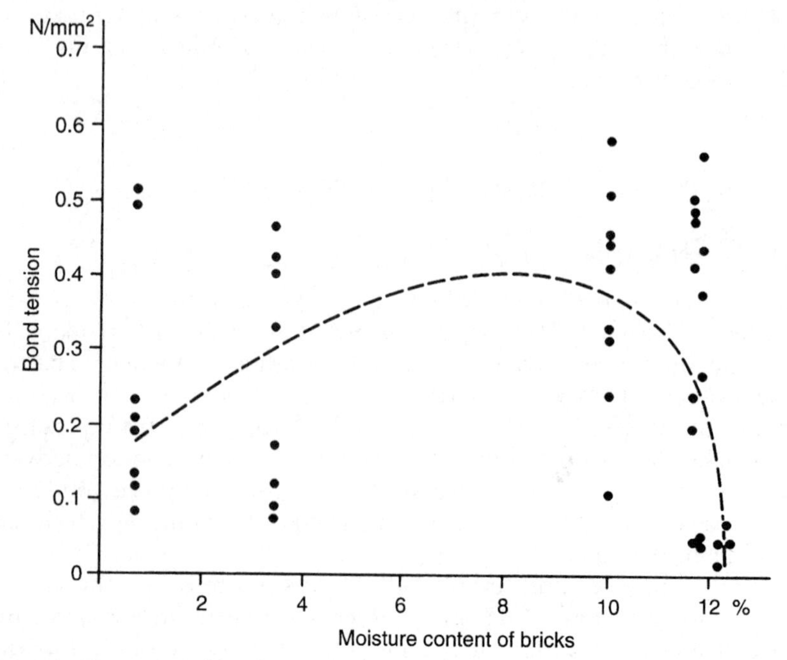

Fig. 3.4 Variation of brick–mortar adhesion with moisture content of bricks at time of laying.

If a wall is supported only at its base and top, its lateral resistance will depend on the flexural tensile strength developed across the bed joints. If it is supported also on its vertical edges, lateral resistance will depend also on the flexural strength of the brickwork in the direction at right angles to the bed joints. The strength in this direction is typically about three times as great as across the bed joints. If the brick–mortar adhesion is good, the bending strength parallel to the bed joint direction will be limited by the flexural tensile strength of the units. If the adhesion is poor, this strength will be limited mainly by the shear strength of the unit–mortar interface in the bed joints.

The flexural tensile strength of clay brickwork ranges from about 2.0 to 0.8 N/mm^2 in the stronger direction, the strength in bending across the bed joints being about one-third of this. As in the case of direct tension, the strength developed is dependent on the absorption characteristics of the bricks and also on the type of mortar used. Calcium silicate brickwork and concrete blockwork have rather lower flexural tensile strength than clay brickwork, that of concrete blockwork depending on the compressive strength of the unit and the thickness of the wall.

3.5 STRESS–STRAIN PROPERTIES OF MASONRY

Masonry is generally treated as a linearly elastic material, although tests indicate that the stress–strain relationship is approximately parabolic, as shown in Fig. 3.5. Under service conditions masonry is stressed only up to a fraction of its ultimate load, and therefore the assumption of a linear stress–strain curve is acceptable for the calculation of normal structural deformations.

Various formulae have been suggested for the determination of Young's modulus. This parameter is, however, rather variable even for nominally identical specimens, and as an approximation, it may be assumed that

$$E = 700\,\sigma_c' \tag{3.3}$$

where σ_c' is the crushing strength of the masonry. This value will apply up to about 75% of the ultimate strength.

For estimating long-term deformations a reduced value of E should be used, in the region of one-half to one-third of that given by equation (3.3).

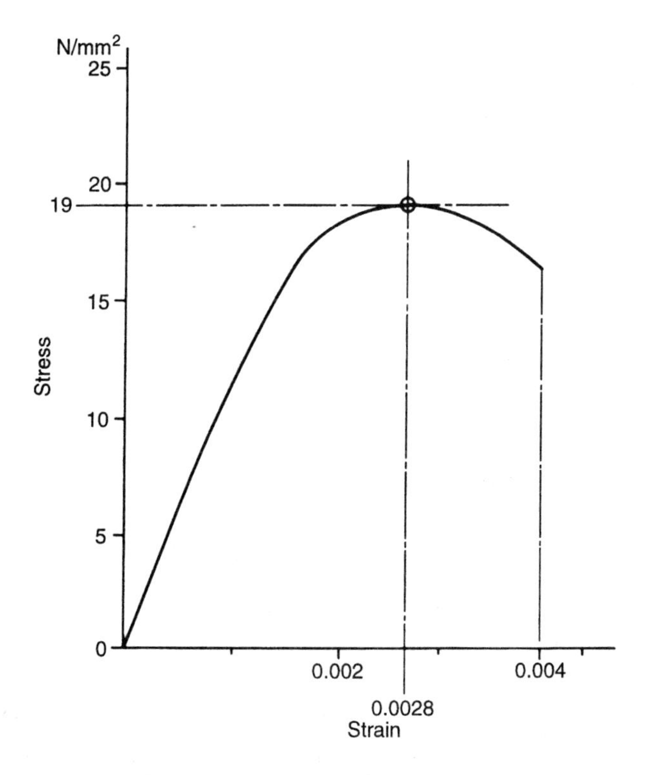

Fig. 3.5 Typical stress–strain curve for brick masonry.

3.6 EFFECTS OF WORKMANSHIP ON MASONRY STRENGTH

Masonry has a very long tradition of building by craftsmen, without engineering supervision of the kind applied to reinforced concrete construction. Consequently, it is frequently regarded with some suspicion as a structural material and carries very much higher safety factors than concrete. There is, of course, some justification for this, in that, if supervision is non-existent, any structural element, whether of masonry or concrete, will be of uncertain strength. If, on the other hand, the same level of supervision is applied to masonry as is customarily required for concrete, masonry will be quite as reliable as concrete. It is therefore important for engineers designing and constructing in masonry to have an appreciation of the workmanship factors which are significant in developing a specified strength. This information has been obtained by carrying out tests on walls which have had known defects built into them and comparing the results with corresponding tests on walls without defects. In practice, these defects will be present to some extent and, in unsatisfactory work, a combination of them could result in a wall being only half as strong in compression as it should be. Such a wall, however, would be obviously badly built and would be so far outside any reasonable specification as to be quite unacceptable.

It is, of course, very much better for masonry to be properly built in the first instance, and time spent by the engineer explaining the importance of the points outlined below to the brick- or blocklayer and his immediate supervisor will be time well spent.

3.6.1 Workmanship defects in brickwork

(a) Failure to fill bed joints

It is essential that the bed joints in brickwork should be completely filled. Gaps in the mortar bed can result simply from carelessness or haste or from a practice known as 'furrowing', which means that the bricklayer makes a gap with his trowel in the middle of the mortar bed parallel to the face of the wall. Tests show that incompletely filled bed joints can reduce the strength of brickwork by as much as 33%.

Failure to fill the vertical joints has been found to have very little effect on the compressive strength of brickwork but does reduce the flexural resistance. Also, unfilled perpendicular joints are undesirable from the point of view of weather exclusion and sound insulation as well as being indicative of careless workmanship generally.

(b) Bed joints of excessive thickness

It was pointed out in discussing the compressive strength of brickwork that increase in joint thickness has the effect of reducing masonry strength because it generates higher lateral tensile stresses in the bricks than would be the case with thin joints. Thus, bed joints of 16–19 mm thickness will result in a reduction of compressive strength of up to 30% as compared with 10 mm thick joints.

(c) Deviation from verticality or alignment

A wall which is built out of plumb, which is bowed or which is out of alignment with the wall in the storey above or below will give rise to eccentric loading and consequent reduction in strength. Thus a wall containing a defect of this type of 12–20 mm will be some 13–15% weaker than one which does not.

(d) Exposure to adverse weather after laying

Newly laid brickwork should be protected from excessive heat or freezing conditions until the mortar has been cured. Excessive loss of moisture by evaporation or exposure to hot weather may prevent complete hydration of the cement and consequent failure to develop the normal strength of the mortar. The strength of a wall may be reduced by 10% as a result. Freezing can cause displacement of a wall from the vertical with corresponding reduction in strength. Proper curing can be achieved by covering the work with polythene sheets, and in cold weather it may also be necessary to heat the materials if bricklaying has to be carried out in freezing conditions.

(e) Failure to adjust suction of bricks

A rather more subtle defect can arise if slender walls have to be built using highly absorptive bricks. The reason for this is illustrated in Fig. 3.6, which suggests how a bed joint may become 'pillow' shaped if the bricks above it are slightly rocked as they are laid. If water has been removed from the mortar by the suction of the bricks, it may have become too dry for it to recover its originally flat shape. The resulting wall will obviously lack stability as a result of the convex shape of the mortar bed and may be as much as 50% weaker than should be expected from consideration of the brick strength and mortar mix. The remedy is to wet the bricks before laying so as to reduce their suction rate below $2\,kg/m^2/min$, and a proportion of lime in the mortar mix will help to retain water in it against the suction of the bricks.

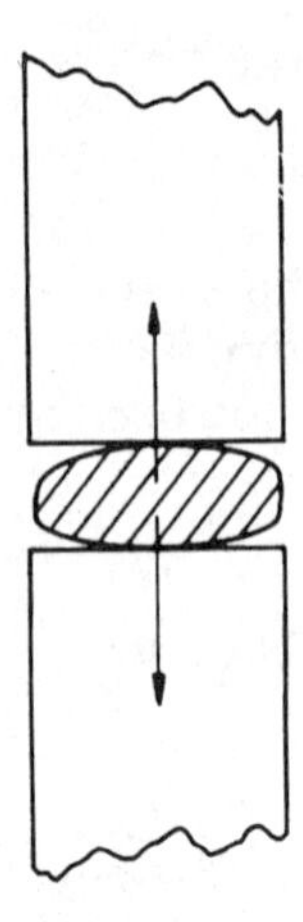

Fig. 3.6 Effect of moisture absorption from mortar bed. Movement of bricks after laying results in 'pillow' shaped mortar bed.

(f) Incorrect proportioning and mixing of mortar

The effect of mortar strength on the strength of masonry may be judged from Fig. 3.1 from which it may be seen with bricks having a crushing strength of $30\,N/mm^2$ that reducing the mortar strength from $11\,N/mm^2$ to $4.5\,N/mm^2$ may be expected to reduce the brickwork strength from $14\,N/mm^2$ to $11\,N/mm^2$. This corresponds to a change in mortar mix from 1:3 cement:sand to 1:4.5 or about 30% too little cement in the mix. A reduction in mortar strength could also result from a relatively high water/cement ratio whilst still producing a workable mix. It is therefore important to see that the specification for mortar strength is adhered to although there is an inherent degree of tolerance sufficient to accommodate small errors in proportioning and mixing the mortar. The use of unsuitable or an excessive amount of plasticizer in place of lime will produce a porous and possibly weak mortar and has to be guarded against.

3.6.2 Workmanship defects in concrete blockwork

Most of the studies on the effect on the compressive strength of masonry, on which the above discussion is based, have been carried out on clay brickwork walls. Some of the factors described, however, apply also to concrete blockwork including the need to fill bed joints and for walls to be built accurately in terms of verticality, planeness and alignment. Excessively thick joints are less likely to be significant in blockwork but the need to meet the specified mortar mix or strength is equally important. Protection against adverse weather conditions is again necessary.

4

Codes of practice for structural masonry

4.1 CODES OF PRACTICE: GENERAL

A structural code of practice or standard for masonry brings together essential data on which to base the design of structures in this medium. It contains recommendations for dealing with various aspects of design based on what is generally considered to be good practice at the time of preparing the code. Such a document is not, however, a textbook and does not relieve the designer from the responsibility of acquiring a full understanding of the materials used and of the problems of structural action which are implicit in his or her design. It follows therefore that, in order to use a code of practice satisfactorily, and perhaps even safely, the engineer must make a careful study of its provisions and, as far as possible, their underlying intention. It is not always easy to do this, as codes are written in terms which often conceal the uncertainties of the drafters, and they are seldom accompanied by commentaries which define the basis and limitations of the various clauses.

This chapter is devoted to a general discussion of the British Code of Practice, BS 5628: Parts 1 and 2, which deal respectively with unreinforced and reinforced masonry, and also with ENV 1996-1-1. The latter document covers both unreinforced and reinforced masonry and after a trial period will become Eurocode 6 (EC6). The application of these codes will be discussed in detail in subsequent chapters of this book.

4.2 THE BASIS AND STRUCTURE OF BS 5628: PART 1

The British code is based on limit state principles, superseding an earlier code in permissible stress terms. The code is arranged in the following five sections:

- *Section 1.* General: scope, references, symbols, etc.
- *Section 2.* Materials, components and workmanship
- *Section 3.* Design: objectives and general recommendations
- *Section 4.* Design: detailed considerations
- *Section 5.* Design: accidental damage

There are also four appendices which are not technically part of the code but give additional information on various matters.

4.2.1 Section 1: general

The code covers all forms of masonry including brickwork, blockwork and stone. It is to be noted that the code is based on the assumption that the structural design is to be carried out by a chartered civil or structural engineer or other appropriately qualified person and that the supervision is by suitably qualified persons, although the latter may not always be chartered engineers.

If materials and methods are used that are not referred to by the code, such materials and methods are not ruled out, provided that they achieve the standard of strength and durability required by the code and that they are justified by test.

4.2.2 Section 2: materials, components, symbols, etc.

This section deals with materials, components and workmanship. In general, these should be in accordance with the relevant British Standard (e.g. BS 5628: Part 3; Materials and components, design and workmanship and BS 5390; Stone masonry). Structural units and other masonry materials and components are covered by British Standards, but if used in an unusual way, e.g. bricks laid on stretcher side or on end, appropriate strength tests have to be carried out.

A table in this section of the code (see Table 2.6, section 2.3) sets out requirements for mortar in terms of proportion by volume together with indicative compressive strengths at 28 days for preliminary and site tests. The wording of the paragraph referring to this table seems to suggest that both the mix and the strength requirements have to be satisfied simultaneously – this may give rise to some difficulty as variations in sand grading may require adjustment of the mix to obtain the specified strength. Four mortar mixes are suggested, as previously noted, in terms of volumetric proportion. Grades (i), (ii) and (iii) are the most usual for engineered brickwork. Lower-strength mortars may be more appropriate for concrete blockwork where the unit strength is generally lower and shrinkage and moisture movements greater. Mortar additives, other than calcium chloride, are not ruled out but have to be used with care.

In using different materials in combination, e.g. clay bricks and concrete blocks, it is necessary to exercise considerable care to allow differential movements to take place. Thus the code suggests that more flexible wall ties may be substituted for the normal vertical twist ties in cavity walls in which one leaf is built in brickwork and the other in blockwork.

4.2.3 Sections 3 and 4: design

Sections 3 and 4 contain the main design information, starting with a statement of the basis of design. Unlike its predecessor, CP 111, BS 5628 is based on limit state principles.

It is stated that the primary objective in designing loadbearing masonry members is to ensure an adequate margin of safety against the attainment of the ultimate limit state. In general terms this is achieved by ensuring that

$$\text{design strength} \geqslant \text{design load}$$

As stated in Chapter 1, the term *design load* is defined as follows:

$$\text{design load} = \text{characteristic load} \times \gamma_f$$

where γ_f is a partial safety factor introduced to allow for (a) possible unusual increases in load beyond those considered in deriving the characteristic load, (b) inaccurate assessment of effects of loading and unforeseen stress redistribution within the structure, and (c) variations in dimensional accuracy achieved in construction.

As a matter of convenience, the γ_f values have (see Table 4.1) been taken in this code to be, with minor differences, the same as in the British code for structural concrete, CP 110: 1971. The effects allowed for by (b) and (c) above may or may not be the same for masonry and concrete. For example, structural analysis methods normally used for the design of concrete structures are considerably more refined than those used for masonry structures. Dimensional accuracy is related to the degree of supervision applied to site construction, which is again normally better for concrete than for masonry. There is, however, no reason why more accurate design methods and better site supervision should not be applied to masonry construction, and as will be seen presently the latter is taken into account in BS 5628 but by adjusting the material partial safety factor γ_m rather than γ_f.

As explained in Chapter 1, characteristic loads are defined theoretically as those which will not be exceeded in 95% of instances of their application, but as the information necessary to define loads on a statistical basis is seldom available, conventional values are adopted from relevant codes of practice, in the present case from the British Standard Codes of Practice CP 3, Chapter V.

Table 4.1 Partial safety factors in BS 5628

(A) *Partial safety factors for loads* (γ_f)

(a) *Dead and imposed load*

$$\text{design dead load} = 0.9\,G_k \text{ or } 1.4\,G_k$$
$$\text{design imposed load} = 1.6\,Q_k$$

(b) *Dead and wind load*

$$\text{design dead load} = 0.9\,G_k \text{ or } 1.4\,G_k$$
$$\text{design wind load} = 1.4\,W_k \text{ or } 0.015\,G_k \quad \text{(whichever is the larger)}$$

In the particular case of free-standing walls and laterally loaded wall panels, whose removal should in no way affect the stability of the remaining structure, γ_f applied on the wind load may be taken as 1.2.

(c) *Dead, imposed and wind load*

$$\text{design dead load} = 1.2\,G_k$$
$$\text{design imposed load} = 1.2\,Q_k$$
$$\text{design wind load} = 1.2\,W_k \text{ or } 0.015\,G_k \quad \text{(whichever is the larger)}$$

(d) *Accidental damage*

$$\text{design dead load} = 0.95\,G_k \text{ or } 1.05\,G_k$$

design imposed load $= 0.35\,Q_k$ (except that, in the case of buildings used predominantly for storage, or where the imposed load is of a permanent nature, $1.05\,Q_k$ should be used)

$$\text{design wind load} = 0.35\,W_k$$

Here G_k is the characteristic dead load, Q_k is the characteristic imposed load, W_k is the characteristic wind load, and the numerical values are the appropriate γ_f factors.

(B) *Partial safety factors for materials* (γ_m)

		Category of construction control	
		Special	*Normal*
Category of manufacturing Control	Special	2.5	3.1
	Normal	2.8	3.5

Different values of γ_f are associated with the various loading cases. Reduced values are specified for accidental damage.

Turning now to the other side of the limit state equation, the term *design strength* is defined as:

$$\text{design strength} = \text{characteristic strength}/\gamma_m$$

Values of the material partial safety factor γ_m were established by the Code Drafting Committee. In theory this could have been done by statistical calculations – if the relevant parameters for loads and materials had been known and the desired level of safety (i.e. acceptable probability of failure) had been specified. However, these quantities were not known and the first approach to the problem was to try to arrive at a situation whereby the new code would, in a given case, give walls of the same thickness and material strength as in the old one. The most obvious procedure was therefore to split the global safety factor of about 5 implied in the permissible state code into partial safety factors relating to loads (γ_f) and material strength (γ_m). As the γ_f values were taken from CP 110 this would seem to be a fairly straightforward procedure. However, the situation is more complicated than this – for example, there are different partial safety factors for different categories of load effect; and in limit state design, partial safety factors are applied to characteristic strengths which do not exist in the permissible stress code. Thus more detailed consideration was necessary, and reference was made to the theoretical evaluation of safety factors by statistical analysis. These calculations did not lead directly to the values given in the code but they provided a reference framework whereby the γ_m values selected could be checked. Thus, it was verified that the proposed values were consistent with realistic estimates of variability of materials and that the highest and lowest values of γ_m applying, respectively, to unsupervised and closely supervised work should result in about the same level of safety. It should be emphasized that, although a considerable degree of judgement went into the selection of the γ_m values, they are not entirely arbitrary and reflect what is known from literally thousands of tests on masonry walls.

The values arrived at are set out in Table 4 of the code and are shown in Table 4.1. There are other partial safety factors for *shear* and for *ties*. For *accidental damage* the relevant γ_m values are halved.

It was considered reasonable that the principal partial safety factors for materials in compression should be graded to take into account differences in manufacturing control of bricks and of site supervision. There is therefore a benefit of about 10% for using bricks satisfying the requirement of 'special' category of manufacture and of about 20% for meeting this category of construction control. The effect of adopting both measures is to reduce γ_m by approximately 30%, i.e. from 3.5 to 2.5.

The requirements for 'special' category of manufacturing control are quite specific and are set out in the code. The definition of 'special' category of construction control is rather more difficult to define, but it is stated in Section 1 of the code that 'the execution of the work is carried out under the direction of appropriately qualified supervisors', and in Section 2 that '... workmanship used in the construction of loadbearing walls should comply with the appropriate clause in BS 5628: Part 3 ...'. Taken together these provisions must be met for 'normal category' of construction control. 'Special category' includes these requirements and in addition requires that the designer should ensure that the work in fact conforms to them and to any additional requirements which may be prescribed.

The code also calls for compressive strength tests on the mortar to be used in order to meet the requirements of 'special' category of construction control.

Characteristic strength is again defined statistically as the strength to be expected in 95% of tests on samples of the material being used. There are greater possibilities of determining characteristic strengths on a statistical basis as compared with loads, but again, for convenience, conventional values for characteristic compressive strength are adopted in BS 5628, in terms of brick strength and mortar strength. This information is presented graphically in Fig. 4.1. Similarly, characteristic flexural and shear strengths are from test results but not on a strictly statistical basis. These are shown in Table 4.2.

A very important paragraph at the beginning of Section 3 of BS 5628 draws attention to the responsibility of the designer to ensure overall stability of the structure, as discussed in Chapter 1 of this book. General considerations of stability are reinforced by the requirement that the structure should be able to resist at any level a horizontal force equal to 1.5% of the characteristic dead load of the structure above the level considered. The danger of divided responsibility for stability is pointed out. Accidents very often result from divided design responsibilities: in one well known case, a large steel building structure collapsed as a result of the main frames having been designed by a consulting engineer and the connections by the steelwork contractor concerned – neither gave proper consideration to the overall stability. Something similar could conceivably happen in a masonry structure if design responsibility for the floors and walls was divided.

The possible effect of accidental damage must also be taken into account in a general way at this stage, although more detailed consideration must be given to this matter as a check on the final design.

Finally, attention is directed to the possible need for temporary supports to walls during construction.

Section 4 is the longest part of the code and provides the data necessary for the design of walls and columns in addition to characteristic strength of materials and partial safety factors.

The basic design of compression members is carried out by calculating their design strength from the formula

$$\frac{\beta b t f_k}{\gamma_m} \tag{4.1}$$

where β is the capacity reduction factor for slenderness and eccentricity,

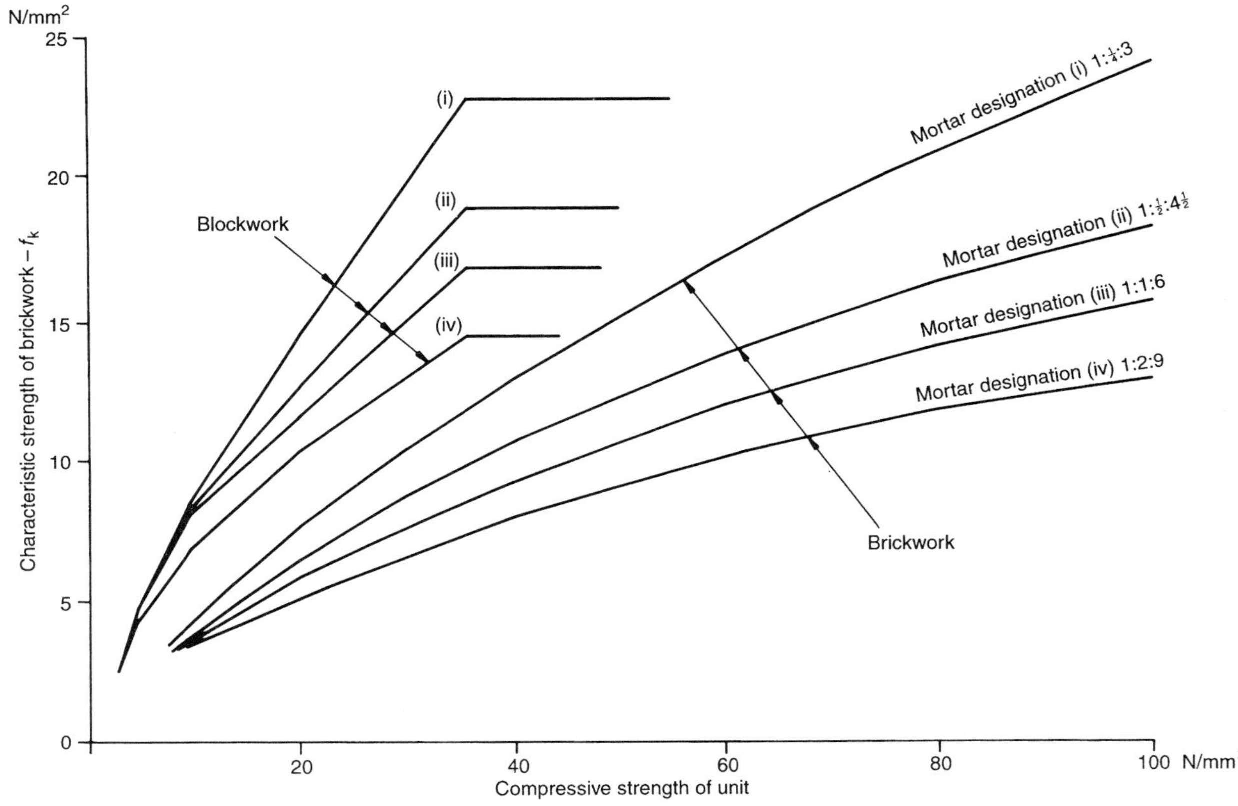

Fig. 4.1 Characteristic strength of brickwork and solid concrete blockwork, where ratio of height to thickness of unit is between 2.0 and 4.0.

Table 4.2 Flexural and shear characteristic strengths in BS 5628 (1992)

(A) *Flexural characteristic strengths for clay brickwork*

Mortar designation	*Plane of failure parallel to bed joints*				*Plane of failure perpendicular to bed joints*			
	(i)	*(ii)*	*(iii)*	*(iv)*	*(i)*	*(ii)*	*(iii)*	*(iv)*
Clay bricks having a water absorption								
less than 7%	0.7	0.5		0.4	2.0	1.5		1.2
between 7% and 12%	0.5	0.4		0.35	1.5	1.1		1.0
over 12%	0.4	0.3		0.25	1.1	0.9		0.8
Calcium silicate bricks	0.3			0.2		0.9		0.6
Concrete bricks	0.3			0.2		0.9		0.6
Concrete blocks (solid or hollow) of compressive strength (N/mm^2)								
2.8 used in walls						0.40		0.4
3.5 of thickness	0.25			0.2		0.45		0.4
7.0 up to 100 mm						0.60		0.5
2.8 used in walls						0.25		0.2
3.5 of thickness	0.15			0.1		0.25		0.2
7.0 up to 250 mm						0.35		0.3
10.5 used in walls						0.75		0.6
14.0 of any	0.25			0.2		0.90		0.7
and over thickness								

(B) *Characteristic shear strengths*

Brickwork built in mortar designation (i) or (ii):

$$f_v = 0.35 + 0.6g_A \quad \text{but not exceeding } 1.75\,N/mm^2$$

Brickwork built in mortar designation (iii) or (iv):

$$f_v = 0.15 + 0.6g_A \quad \text{but not exceeding } 1.4\,N/mm^2$$

Here g_A is the design vertical load per unit area of wall cross-section due to the vertical loads calculated from the appropriate loading condition.

For shear in the vertical plane between brickwork elements bonded together:

(a) for bricks set in mortar designations (i) and (ii)

$$f_v = 0.7\,N/mm^2$$

(b) for bricks set in mortar designations (iii) and (iv)

$$f_v = 0.5\,N/mm^2$$

(c) for dense aggregate solid concrete blocks having a minimum strength of $7\,N/mm^2$ set in mortar designations (i), (ii) or (iii)

$$f_v = 0.35\,N/mm^2$$

b and t are respectively the width and thickness of the member, f_k is the characteristic compressive strength and γ_m is the material partial safety factor.

The capacity reduction factor β has been derived on the assumption that there is a load eccentricity varying from e_x at the top of the wall to zero at the bottom together with an additional eccentricity arising from the lateral deflection related to slenderness. This is neglected if the slenderness ratio (i.e. ratio of effective height to thickness) is less than 6. The additional eccentricity is further assumed to vary from zero at the top and bottom of the wall to a value e_a over the central fifth of the wall height, as indicated in Fig. 4.2. The additional eccentricity is given by an empirical relationship:

$$e_a = t[(1/2400)\,(h_{ef}/t)^2 - 0.015] \tag{4.2}$$

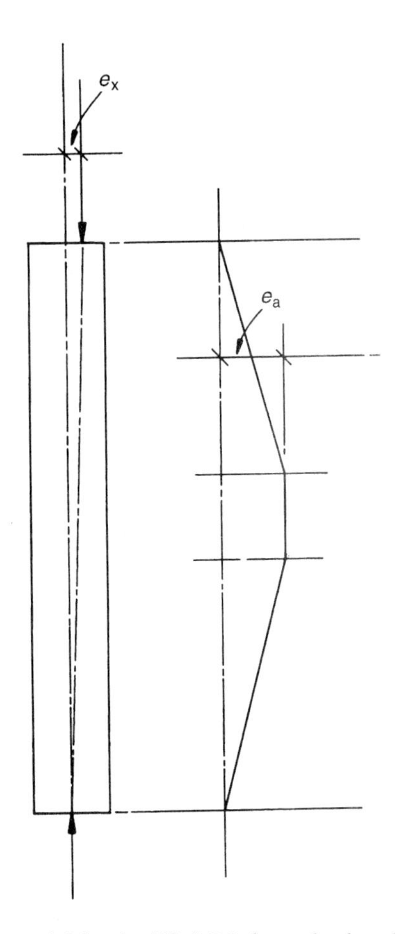

Fig. 4.2 Assumed eccentricities in BS 5628 formula for design vertical load capacity.

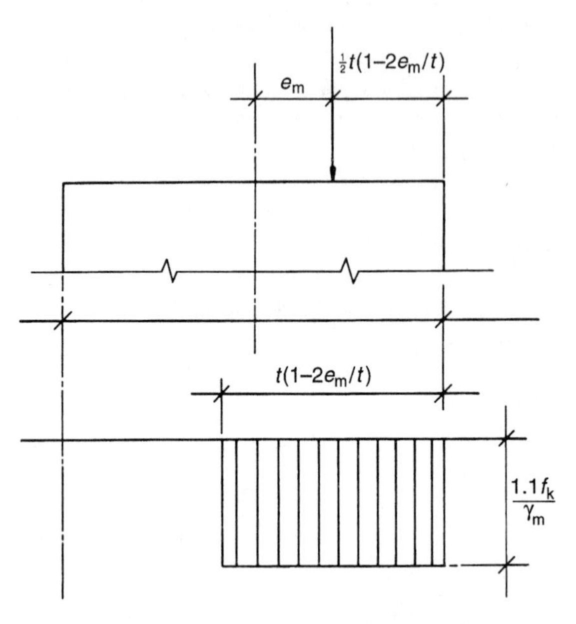

Fig. 4.3 Assumed stress block in BS 5628 formula for design vertical load capacity.

The total eccentricity is then:

$$e_t = 0.6e_x + e_a \tag{4.3}$$

It is possible for e_t to be smaller than e_x, in which case the latter value should be taken as the design eccentricity.

It is next assumed that the load on the wall is resisted by a rectangular stress block with a constant stress of $1.1f_k/\gamma_m$ (the origin of the coefficient 1.1 is not explained in the code but has the effect of making $\beta = 1$ with a minimum eccentricity of $0.05t$).

The width of the stress block, as shown in Fig. 4.3, is

$$t(1 - 2e_m/t) \tag{4.4}$$

and the vertical load capacity of the wall is

$$1.1(1 - 2e_m/t)\,tf_k/\gamma_m \tag{4.5}$$

or

$$\beta t f_k/\gamma_m \tag{4.6}$$

It will be noted that e_m is the larger of e_x and e_t and is to be not less than $0.05t$. If the eccentricity is less than $0.05t$, β is taken as 1.0 up to a slenderness ratio of 8. The resulting capacity reduction factors are shown in Fig. 4.4.

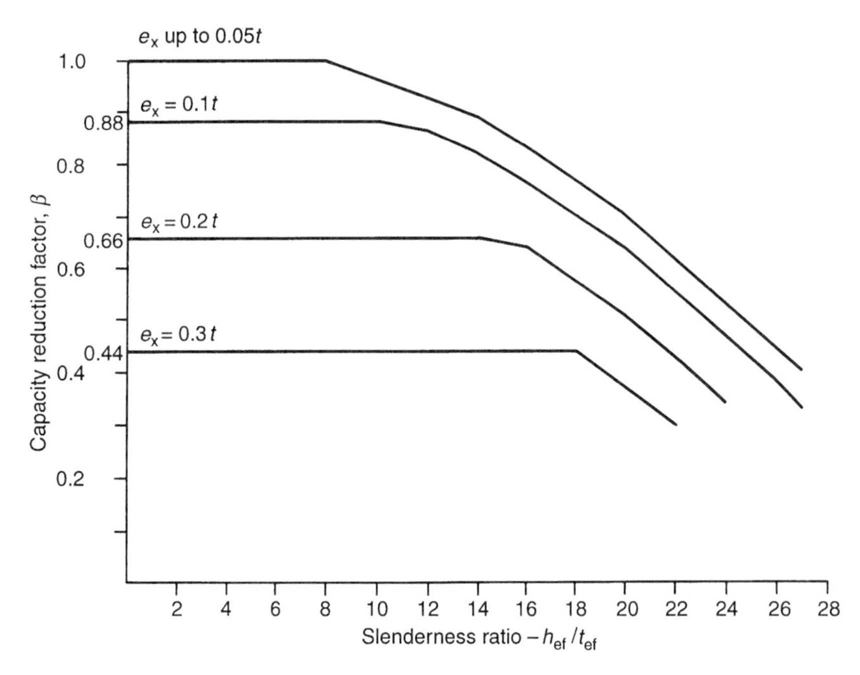

Fig. 4.4 Capacity reduction factor β in BS 5628.

As will be apparent, this method of calculating the capacity reduction factor for slenderness and eccentricity embodies a good number of assumptions, and the simple rules given for estimating the eccentricity at the top of a wall are known to be inaccurate – generally the eccentricities calculated by the code method are very much smaller than experimental values. This, however, may be compensated by the empirical formula used for calculating the additional eccentricity, e_a, and by the other assumptions made in calculating the reduction factor. The final result for loadbearing capacity will be of variable accuracy but, protected as it is by a large safety factor, will result in structures of very adequate strength.

The remaining part of Section 4 deals with concentrated loads and with walls subjected to lateral loading. Concentrated loads on brickwork are associated with beam bearings, and higher stresses are permitted in the vicinity of these loads. The code distinguishes three types of beam bearing, as shown in Fig. 4.5. The local design strength, calculated on a uniform bearing stress, for type 1 bearings is $1.25f_k/\gamma_m$ and for type 2 bearings $1.5f_k/\gamma_m$. Careful inspection of the diagram shown in Fig. 4.5 is necessary to see within which category a particular detail may come, and the logic of the categories is by no means clear. However, it can be seen that under type 1, a slab spanning at right angles to a wall is

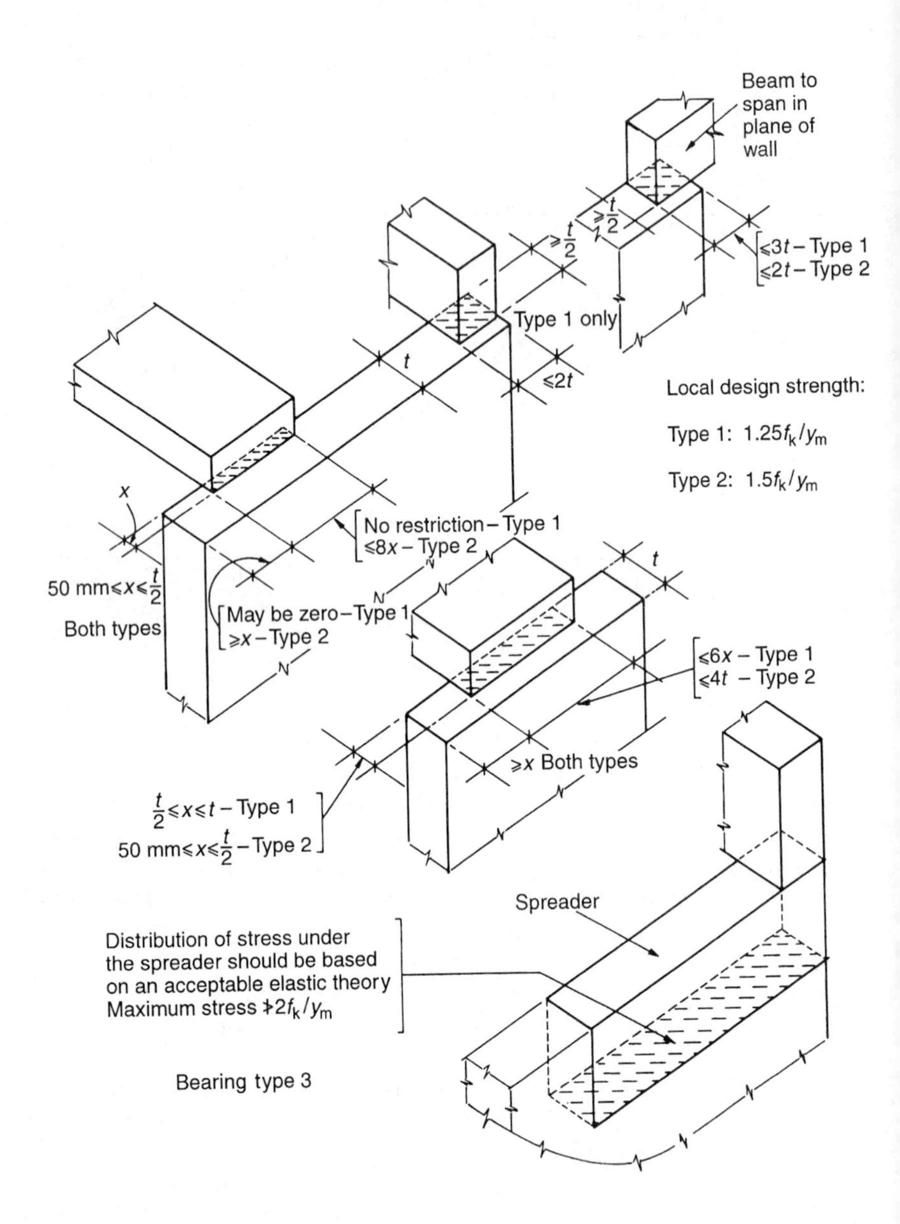

Fig. 4.5 Design stresses in vicinity of various beam and slab bearings according to BS 5628.

allowed a 25% increase in design strength, provided that the bearing width is between 50 mm and half the thickness of the wall. Type 2 includes short beam or slab bearings spanning at right angles to the wall, provided that they are more than the bearing width from the end of the wall. Slabs whose bearing length is between six and eight times their bearing width are included in category 2 and are thus allowed a 50% increase in design strength. A slab resting on the full thickness and width of a wall attracts a 25% increase in design stress provided that it is no longer than six times the wall thickness.

Type 3 bearings envisage the use of a spreader or pad-stone and are permitted a 100% increase in design strength under the spreader. The stress distribution at this location is to be calculated by an acceptable elastic theory.

The accuracy of these rather complicated provisions is uncertain. Test results (Page and Hendry, 1987) for the strength of brickwork under concentrated loading suggest that simpler rules are possible and such have been adopted in EC6 (see subsection 4.4.4 (c)).

The section on laterally loaded walls was based on a programme of experimental research carried out at the laboratories of the British Ceramic Research Association. For non-loadbearing panels the method is to calculate the design moment given by the formula:

$$\alpha W_k \gamma_f L^2 \tag{4.7}$$

where α is a bending moment coefficient, γ_f is the partial safety factor for loads, L is the length of the panel and W_k is the characteristic wind load/unit area. Values of α for a variety of boundary conditions are given in the code. They are numerically the same as obtained by yield line formulae for corresponding boundary conditions.

This moment is compared with the design moment of resistance about an axis perpendicular to the plane of the bed joint, equal to

$$f_{kx} Z / \gamma_m$$

where f_{kx} is the characteristic strength in flexure, γ_m is the partial safety factor for materials and Z is the section modulus.

Obviously everything depends on the successful achievement of f_{kx} on site, and considerable attention must be given to ensuring satisfactory adhesion between bricks and mortar. The best advice that can be given in this respect is to ensure that the bricks are neither kiln-dry nor saturated. Mortar should have as high a water content and retentivity as is consistent with workability. Calcium silicate bricks seem to require particular care in this respect.

Further information is given in this section relating to the lateral resistance of walls with precompression, free-standing walls and retaining walls.

4.2.4 Section 5: accidental damage

The final section of the code deals with the means of meeting statutory obligations in respect of accidental damage. Special measures are called for only in buildings of over four storeys, although it is necessary to ensure that all buildings are sufficiently robust, as discussed in Chapter 1.

For buildings of five storeys and over, three possible approaches are suggested:

1. To consider the removal of one horizontal or vertical member at a time, unless it is capable of withstanding a pressure of $34\,kN/m^2$ in any direction, in which case it may be classed as a 'protected' member.
2. To provide horizontal ties capable of resisting a specified force and then to consider the effect of removing one vertical member at a time (unless 'protected').

In both the above cases the building should remain stable, assuming reduced partial safety factors for loads and materials.

3. To provide horizontal and vertical ties to resist specified forces.

It would appear most practicable to adopt the second of the above methods. The first raises the problem of how a floor could be removed without disrupting the walls as well. In the third option, the effect of vertical ties is largely unknown but in one experiment they were found to promote progressive collapse by pulling out wall panels on floors above and below the site of an explosion. If vertical ties are used it would seem advisable to stagger them from storey to storey so as to avoid this effect.

The treatment of accidental damage is discussed in detail in Chapter 9, and the application of the code provisions to a typical design is given in Chapter 10.

4.3 BS 5628: PART 2 – REINFORCED AND PRESTRESSED MASONRY

Part 2 of BS 5628 is based on the same limit state principles as Part 1 and is set out in seven sections, the first three of which, covering introductory matters, materials and components and design objectives, are generally similar to the corresponding sections of Part 1. Sections 4 and 5 are devoted to the design of reinforced and prestressed masonry, respectively, whilst the remaining two sections give recommendations relating to such matters as durability, fire resistance and site procedures.

4.3.1 Section 1: general

This section lists additional definitions and symbols relating to reinforced and prestressed masonry and notes that the partial safety factors given for this type of construction assume that the special category of construction control specified in Part 1 will apply. If this is not possible in practice, then higher partial safety factors should be used.

4.3.2 Section 2: materials and components

References are given to relevant standards for masonry units, reinforcing steel, wall ties and other items. Requirements for mortar and for concrete infill are stated. Mortar designations (i) and (ii) as in Part 1 are normally to be used but designation (iii) mortar may be used in walls in which bed-joint reinforcement is placed to increase resistance to lateral loading.

A suitable concrete mix for infill in reinforced masonry is given as $1:0-\frac{1}{4}:3:2$, cement:lime:sand:10 mm maximum size aggregate. Other infill mixes for pre- and post-tensioned masonry are quoted with reference to the relevant British Standard, BS 5328, for specifying concrete mixes. Recommendations for admixtures of various kinds are also given.

4.3.3 Section 3: design objectives

As in Part 1, this section sets out the basis of design in limit state terms, including values for characteristic strength of materials and partial safety factors.

In unreinforced brickwork, serviceability limit states rarely require explicit consideration but deflection and cracking may be limiting factors in reinforced or prestressed work. Thus it is suggested that the final deflection of all elements should not exceed length/125 for cantilevers or span/250 for all other elements. To avoid damage to partitions or finishes the part of the deflection taking place after construction should be limited to span/500 or 20 mm and the upward deflection of prestressed members before the application of finishes should not exceed span/300. A general requirement is stated that cracking should not adversely affect appearance or durability of a structure.

Characteristic strengths of brickwork in compression follow Part 1 with an additional clause covering the case in which compressive forces act parallel to the bed faces of the unit. As indicated in section 3.2.6 of the code the characteristic strength of brickwork stressed in this way may have to be determined by test if cellular or perforated bricks are used. The code suggests a lower-bound value of one-third of the normal strength if test data are not available.

Shear strength for brickwork sections reinforced in bed or vertical joints is given as 0.35 N/mm^2. In the case of grouted cavity or similar sections, this value is augmented by 17.5 ρ, where ρ is the steel ratio. To allow for the increased shear strength of beams or cantilever walls where the shear span ratio (a/d) is less than 6, the characteristic shear strength may be increased by a factor $[2.5 - 0.25(a/d)]$ up to a maximum of 1.7 N/mm^2.

Racking shear strength for walls is the same as for unreinforced walls except that in walls in which the main reinforcement is placed within pockets, cores or cavities the characteristic shear strength may be taken as 0.7 N/mm^2, provided that the ratio of height to length does not exceed 1.5.

For prestressed sections, the shear strength is given as $f_v = (0.35 + 0.6g)$ N/mm^2, where g is the design load acting at right angles to the bed joints, including prestressing loads. If, however, the prestressing force acts parallel to the bed joints, $g = 0$ and $f_v = 0.35\,N/mm^2$. These values may again be increased when the shear span ratio is less than 6.

The characteristic tensile strength of various types of reinforcing steel is as shown in Table 2.9.

As it will be necessary in some cases to check deflections of reinforced and prestressed elements, values are given for the elastic moduli of the various materials involved. For brickwork under short-term loading $E = 0.9 f_k\,kN/mm^2$ and for long-term loading $0.45 f_k\,kN/mm^2$ for clay brickwork and $0.3 f_k\,kN/mm^2$ for calcium silicate brickwork. The elastic modulus of concrete infill varies with the cube strength as shown in Table 4.3.

Partial safety factors are generally as in Part 1, but with the addition of ultimate limit state values of 1.5 and 1.15 for bond strength between infill and steel and for steel, respectively. It is assumed that the 'special' category of construction control will normally apply to reinforced and prestressed work.

4.3.4 Section 4: design of reinforced masonry

Section 4 is subdivided into paragraphs dealing with the design of elements subjected to bending, combined vertical loading and bending, axial compressive loading and horizontal forces in their own plane. The principles underlying the design methods and formulae are the same as for reinforced concrete, with suitable modifications to allow for differences in material properties. The formulae given for the design of simply reinforced, rectangular beams allow for flexural failure by yielding of the steel with a cut-off to exclude brittle failures. These principles and related formulae will be discussed in detail in Chapter 10 along with examples of their application.

Table 4.3 Elastic modulus for concrete infill

28-day cube strength (N/mm^2)	20	30	40	50	60
E (kN/mm^2)	24	26	28	30	32

A final subsection of Section 4 gives recommendations for reinforcement details.

4.3.5 Section 5: design of prestressed masonry

The design methods given in this section for prestressed elements are again similar to those which have been developed for prestressed concrete. Calculation of the moment of resistance at the ultimate limit state is to be based on the assumption of linear strain distribution and a rectangular stress block in the compression zone, omitting the tensile strength of the masonry.

Design for the serviceability limit state is provided for by limiting the compressive stresses at transfer of the prestressing force and after all losses have occurred. Calculation of tendon forces must allow for loss of prestress resulting from a variety of causes and information is provided on which to base these estimates. Finally, a short subsection gives rules for detailing anchorages and tendons.

Experience in the use of prestressed brickwork on which the code has to be based is more limited than for reinforced brickwork and therefore the provisions of this part of the document are necessarily less detailed and in some cases rather tentative.

4.3.6 Sections 6 and 7: other design considerations and work on site

Section 6 of the code deals with the important matter of durability and, specifically, with the selection of material for avoidance of corrosion of reinforcement in various conditions of exposure, as defined in Part 3 of BS 5628. Where carbon steel is used, minimum concrete cover for these exposure conditions is specified.

Section 7, dealing with work on site, also refers to Part 3 of BS 5628 and gives additional guidance on a number of matters specifically relating to reinforced and prestressed work, such as the procedures to be adopted in filling cavities in grouted cavity, Quetta bond or similar forms of construction. It is again stated that the special category of construction control should be specified for this type of work.

4.4 DESCRIPTION OF EUROCODE 6 PART 1-1 (ENV 1996-1-1: 1995)

Eurocode 6 is one of a group of standards for structural design being issued by the Commission of the European Communities. It was published in draft form in 1988 and, following a lengthy process of comment and review, the first part was issued in 1995 as a 'pre-standard' or ENV under the title *Part 1-1: General rules for buildings. Rules for reinforced and unreinforced masonry*. Following a trial period of use on a voluntary basis, the document will be reissued as a Eurocode, taking account of any amendments shown to be necessary. Other parts of EC6 dealing with special aspects of masonry design are being prepared or are planned. Eurocodes for the various structural materials all rely on EC1 for the specification of the basis of design and actions on structures.

EC6 Part 1-1 is laid out in the following six sections:

- *Section 1*. General
- *Section 2*. Basis of design
- *Section 3*. Materials
- *Section 4*. Design of masonry
- *Section 5*. Structural detailing
- *Section 6*. Construction

The clauses in ENV 1996-1-1 are of two categories, namely, 'Principles', designated by the letter P, and 'Application rules'. In general, no alternatives are permitted to the principles but it is permissible to use alternatives to the application rules, provided that they accord with the principles.

A further point to be noted in using the code is that many of the values for material strengths and partial safety factors are shown 'boxed'. This is because national authorities have responsibility for matters affecting safety and may, in an accompanying National Application Document, specify values which differ from the indicative figures shown in the ENV.

The following paragraphs give a summary of the content of ENV 1996-1-1 but careful study of its lengthy and complex provisions are necessary before attempting to use it in design.

4.4.1 Section 1: general

The scope of EC6 extends to the design of unreinforced, reinforced and prestressed masonry and also to what is called 'confined' masonry, which is defined as masonry enclosed on all four sides within a reinforced concrete or reinforced masonry frame (steel frames are not mentioned).

It is assumed that structures are designed and built by appropriately qualified and experienced personnel and that adequate supervision exists in relation to unit manufacture and on site. Materials have to meet the requirements of the relevant European standard (EN). It is further assumed that the structure will be adequately maintained and used in accordance with the design brief.

The section contains an extensive list of definitions including a multilingual list of equivalent terms, essential in a document which is to be used throughout the European Community. It concludes with a schedule of the numerous symbols used in the text.

4.4.2 Section 2: basis of design

The code is based on limit state principles and in this section are defined the design situations which have to be considered. Actions, which include loads and imposed deformations (for example arising from thermal effects or settlement), are obtained from EC1 (ENV 1991) or other approved sources. Indicative values for partial safety factors for actions are as shown in Table 4.4.

Application of these safety factors requires a distinction to be made between actions which are permanent or which vary with time or which may change in position or extent. Combinations of actions require the application of coefficients to the various actions concerned and general formulae for such combinations are given. Values of the combination coefficients are provided in ENV 1991, but for building structures the following formulae may be used in conjunction with the partial safety factors for the ultimate limit state shown in Table 4.4.

Considering the most unfavourable variable action:

$$\sum \gamma_{G,j} G_{k,j} + 1.5 Q_{k,l} \tag{4.8}$$

Considering all unfavourable variable actions:

$$\sum \gamma_{G,j} G_{k,j} + 1.35 \sum_{i \geqslant l} Q_{k,i} \tag{4.9}$$

Table 4.4 Partial safety factors for actions in building structures for persistent and transient design situations

	Permanent actions, γ_G[a]	*Variable actions,* γ_Q		*Prestressing,* γ_p
		One with its characteristic value	*Others with their combination value*	
Favourable	1.0	0	0	0.9
Unfavourable	1.35	1.5	1.35	1.2

[a]See also paragraph 2.3.3.1(3) of EC6.

whichever gives the larger value, where $\gamma_{G,j}$ is the partial safety factor for permanent actions, $G_{k,j}$ is the characteristic value of permanent actions and $Q_{k,l}$ and $Q_{k,i}$ are respectively, the characteristic values of the most and of the other variable actions considered.

Partial safety factors for material properties are given, as in Table 4.5. These are applied as appropriate to the characteristic material strengths to give design strengths.

4.4.3 Section 3: materials

(a) Units and mortar

This section starts by defining masonry units, first in terms of relevant European standards and then by categories which reflect quality control in manufacture and also with reference to the volume and area of holes which there may be in a unit.

Mortars are classified according to their compressive strength (determined according to EN 1015-11) or by mix proportions. If specified by strength the classification is indicated by the letter M followed by the compressive strength in N/mm^2.

Requirements are also set out for unit and mortar durability and for the properties of infill concrete and reinforcing steel.

Table 4.5 Partial safety factors for material properties, γ_M (EC6)

γ_M			*Category of execution* [a]		
			A	*B*	*C*
Masonry[b]	Category of manufacturing control of masonry units[c]	I	1.7	2.2	2.7
		II	2.0	2.5	3.0
Anchorage and tensile and compressive resistance of wall ties and straps			2.5	2.5	2.5
Anchorage bond of reinforcing steel			1.7	2.2	–
Steel (referred to as γ_S)			1.15	1.15	–

[a] See section 6.9 of EC6.
[b] The value of γ_M for concrete infill should be taken as that appropriate to the category of manufacturing control of the masonry units in the location where the infill is being used.
[c] See section 3.1 of EC6.

(b) Characteristic compressive strength

Three methods for determining the compressive strength of unreinforced masonry are set out. The first, designated as a principle, states that this shall be determined from the results of tests on masonry. A subsidiary note indicates that such results may be available nationally or from tests carried out for the project. The second method, not designated as a principle, appears to be an elaboration of the first in specifying that tests should be carried out according to EN 1052-1 or from an evaluation of test data in a similar way to that prescribed in the third method.

According to the latter, which may be used in the absence of specific test results or national data, a formula is given, for masonry built with general-purpose mortar, relating unit and mortar strengths to masonry characteristic strength with adjustment for unit proportions and wall characteristics. This formula is as follows:

$$f_k = K f_b^{0.65} f_m^{0.25} \text{ N/mm}^2 \tag{4.10}$$

where f_k is the characteristic masonry strength, f_b is the normalized unit compressive strength, f_m is the specified compressive strength of mortar and K is a constant depending on the construction.

The normalized unit compressive strength is introduced in an attempt to make the formula apply to units of different geometric proportions by making f_b in the formula equivalent to the strength of a 100 mm cube. This is achieved by the use of the factor δ in Table 4.6.

Values of K range from 0.6 to 0.4. The higher value applies to masonry in which the wall thickness is equal to the width of the unit and which in this case is of category I in terms of quality control in manufacture. The lower value applies to masonry in which there is a longitudinal joint in the thickness of the wall, and built of category 2b or 3 units. Intermediate values are given for other cases.

Other formulae are suggested for masonry built with thin-layer or lightweight mortar and for shell-bedded, hollow block masonry.

Table 4.6 Values of factor δ^a (EC6)

Height of unit (mm)	*Least horizontal dimension of unit* (mm)				
	50	100	150	200	250 *or greater*
50	0.85	0.75	0.70	–	–
65	0.95	0.85	0.75	0.70	0.65
100	1.15	1.00	0.90	0.80	0.75
150	1.30	1.20	1.10	1.00	0.95
200	1.45	1.35	1.25	1.15	1.10
250 or greater	1.55	1.45	1.35	1.25	1.15

[a] Linear interpolation is permitted.

It is likely that National Application Documents will prescribe masonry compressive strengths in accordance with experience in the country for which each is issued.

(c) Characteristic shear strength of unreinforced masonry

The characteristic shear strength of unreinforced masonry is to be determined in a similar way to compressive strength, that is on the basis of tests, the results of which may be available nationally, or from tests conducted according to European standards or from the following formulae:

$$f_{vk} = f_{vk0} + 0.4\,\sigma_d \qquad (4.11)$$

or

$$f_{vk} = 0.065 f_b \text{ but not less than } f_{vk0}$$

or

$$f_{vk} = \text{limiting value given in Table 4.7}$$

where f_{vk0} is the shear strength under zero compressive stress or, for general-purpose mortars, the value shown in Table 4.7, σ_d is the design compressive stress normal to the shear stress and f_b is the normalized compressive strength of the units.

Where national data are not available or where tests in accordance with European standards have not been carried out, the value of f_{vk0} should be taken as $0.1\,N/mm^2$.

Other values are given for masonry in which the vertical joints have not been filled and for shell-bedded blockwork.

(d) Flexural strength of unreinforced masonry

The flexural strength of unreinforced masonry is again to be determined by tests or on the basis of national data. Flexural strength is only to be relied upon in the design of walls for resistance to transient actions, such as wind loads.

No values are suggested and it is assumed that these will be specified in National Application Documents.

(e) Anchorage bond strength of reinforcement in infill and in mortar

Values are quoted for anchorage bond strength for plain and high-bond carbon steel and stainless steel embedded in infill concrete and in mortar. These values are higher where the infill concrete is confined within masonry units.

Table 4.7 Values of f_{vk0} and limiting values of f_{vk} for general-purpose mortar (EC6)[a]

Masonry unit	*Mortar*	f_{vk0} (N/mm^2)	*Limiting* f_{vk} (N/mm^2)
Group 1 clay units	M10 to M20	0.3	1.7
	M2.5 to M9	0.2	1.5
	M1 to M2	0.1	1.2
Group 1 units other than clay and natural stone	M10 to M20	0.2	1.7
	M2.5 to M9	0.15	1.5
	M1 to M2	0.1	1.2
Group 1 natural stone units	M2.5 to M9	0.15	1.0
	M1 to M2	0.1	1.0
Group 2a clay units	M10 to M20	0.3	The lesser of longitudinal compressive or 1.4
	M2.5 to M9	0.2	1.2
	M1 to M2	0.1	1.0
Group 2a and Group 2b units other than clay and Group 2b clay units	M10 to M20	0.2	1.4
	M2.5 to M9	0.15	1.2
	M1 to M2	0.1	1.0
Group 3 clay units	M10 to M20	0.3	No limits other than given by equation (3.4) of EC6
	M2.5 to M9	0.2	
	M1 to M2	0.1	

[a]For Group 2a and 2b masonry units, the longitudinal compressive strength of the units is taken to be the measured strength, with δ taken to be not greater than 1.0. When the longitudinal compressive strength can be expected to be greater than $0.15f_b$, by consideration of the pattern of holes, tests are not necessary.

(f) Deformation properties of masonry

It is stated that the stress–strain relationship for masonry is parabolic in form but may for design purposes be assumed as an approximation to be rectangular or parabolic–rectangular. The latter is a borrowing from reinforced concrete practice and may not be applicable to all kinds of masonry.

The modulus of elasticity to be assumed is the secant modulus at the serviceability limit, i.e. at one-third of the maximum load. Where the results of tests in accordance with the relevant European standard are not available E under service conditions and for use in structural analysis may be taken as $1000f_k$. It is further recommended that the E value should be multiplied by a factor of 0.6 when used in determining the serviceability limit state. A reduced E value is also to be adopted in relation to long-term loads. This may be estimated with reference to creep data.

In the absence of more precise data, the shear modulus may be assumed to be 40% of E.

(g) Creep, shrinkage and thermal expansion

A table is provided of approximate values to be used in the calculation of creep, shrinkage and thermal effects. However, as may be seen from Table 4.8 these values are given in terms of rather wide ranges so that it is difficult to apply them in particular cases in the absence of test results for the materials being used.

4.4.4 Section 4: design of masonry

(a) General stability

Initial provisions of this section call for overall stability of the structure to be considered. The plan layout of the building and the interconnection

Table 4.8 Deformation properties of unreinforced masonry made with general-purpose mortar (EC6).

Type of masonry unit	*Final creep coefficient* [a], ϕ_∞		*Final moisture expansion or shrinkage* [b] (mm/m)		*Coefficient of thermal expansion* (10^{-6}/K)	
	Range	*Design value*	*Range*	*Design value*	*Range*	*Design value*
Clay	0.5 to 1.5	1.0	−0.2 to +1.0	— [c]	4 to 8	6
Calcium silicate	1.0 to 2.0	1.5	−0.4 to −0.1	−0.2	7 to 11	9
Dense aggregate concrete and manufactured stone	1.0 to 2.0	1.5	−0.6 to −0.1	−0.2	6 to 12	10
Lightweight aggregate concrete	1.0 to 3.0	2.0	−1.0 to −0.2	−0.4 [d] −0.2 [e]	8 to 12	10
Autoclaved aggregate concrete	1.0 to 2.5	1.5	−0.4 to +0.2	−0.2	7 to 9	8
Natural stone	— [f]	0	−0.4 to +0.7	−0.1	3 to 12	7

[a] The final creep coefficient $\phi_\infty = \varepsilon_{c\infty}/\varepsilon_{el}$, where $\varepsilon_{c\infty}$ is the final creep strain and $\varepsilon_{el} = \sigma/E$.
[b] Where the final value of moisture expansion or shrinkage is shown 'minus' it indicates shortening and where 'plus' it indicates extension.
[c] Values depend upon the type of material concerned and a single design value cannot be given.
[d] Value given is for pumice and expanded clay aggregates.
[e] Value given is for lightweight aggregates other than pumice or expanded clay.
[f] Values are normally very low.

of elements must be such as to prevent sway. The possible effects of imperfections should be allowed for by assuming that the structure is inclined at an angle of $1/100\sqrt{h_{tot}})$ to the vertical where h_{tot} is the total height of the building. One designer must, unambiguously, be responsible for ensuring overall stability.

(b) Accidental damage

Buildings are required to be designed in such a way that there is a 'reasonable probability' that they will not collapse catastrophically under the effect of misuse or accident and that the extent of damage will not be disproportionate to the cause. This is to be achieved by considering the removal of essential loadbearing members or designing them to resist the effects of accidental actions. However, no specific rules relating to these requirements are given.

(c) Design of structural members

The design of members has to be such that no damage is caused to facings, finishes, etc., but it may be assumed that the serviceability limit state is satisfied if the ultimate limit state is verified. It is also required that the stability of the structure or of individual walls is ensured during construction.

Subject to detailed provisions relating to the type of construction, the design vertical load resistance per unit length, N_{Rd}, of an unreinforced masonry wall is calculated from the following expression:

$$N_{Rd} = \Phi_{i,m} t f_k / \gamma_m \tag{4.12}$$

where $\Phi_{i,m}$ is a capacity reduction factor allowing for the effects of slenderness and eccentricity (Φ_i applies to the top and bottom of the wall; Φ_m applies to the mid-height and is obtained from the graph shown in Fig. 4.6), t is the thickness of the wall, f_k is the characteristic compressive strength of the masonry and γ_m is the partial safety factor for the material.

The capacity reduction factor Φ_i is given by:

$$\Phi_i = 1 - 2e_i/t \tag{4.13}$$

where e_i is the eccentricity at the top or bottom of the wall calculated from

$$e_i = M_i/N_i + e_{hi} + e_a > 0.05t \tag{4.14}$$

where M_i and N_i are respectively the design bending moment and vertical load at the top or bottom of the wall and e_{hi} and e_a are eccentricities

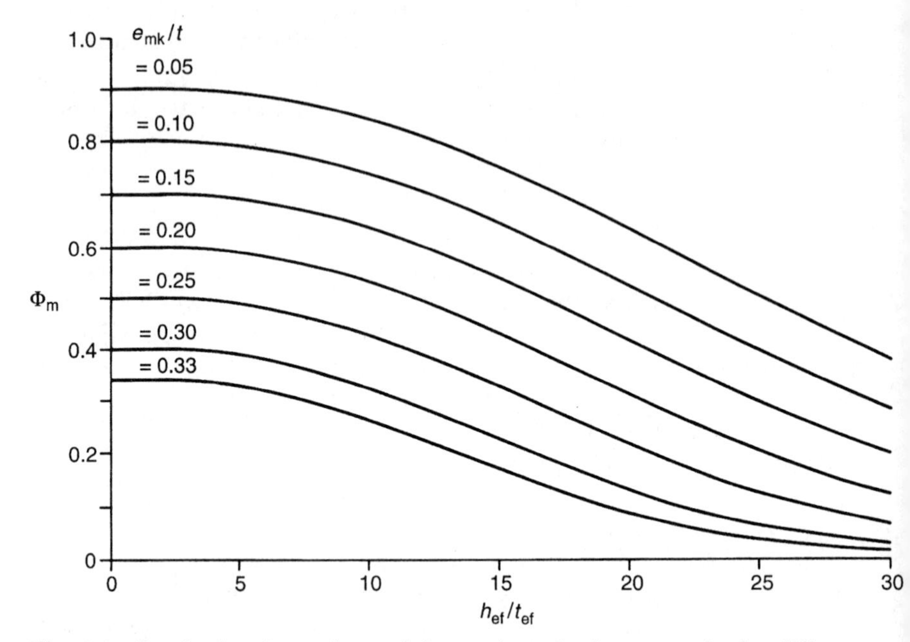

Fig. 4.6 Graph showing values of Φ_m against slenderness ratio for different eccentricities.

resulting from lateral loads and construction inaccuracies, respectively. The recommended value of e_a for average level of construction is $h_{ef}/450$.

The basis of the capacity reduction factor is not stated but is known to derive from a complex theoretical solution originally developed for plain concrete sections (Kukulski and Lugez, 1966). The eccentricity at mid-height, e_{mk}, used in calculating Φ_m is given by

$$e_{mk} = e_m + e_k \geqslant 0.05\,t \tag{4.15}$$

where e_m, the structural eccentricity, is obtained from

$$e_m = M_m/N_m + e_{hm} \pm e_k \tag{4.16}$$

where M_m and N_m are respectively the greatest bending moments and vertical loads within the middle one-fifth of the height of the wall. The eccentricity e_k is an allowance for creep:

$$e_k = 0.002\,\Phi_\infty\,(h_{ef}/t_{ef})\,(te_m)^{1/2} \tag{4.17}$$

being a final creep coefficient (see Table 4.8) equal to zero for walls built of clay and natural stone units.

Rules are given for the assessment of the effective height of a wall. In general, walls restrained top and bottom by reinforced concrete slabs are assumed to have an effective height of 0.75 × actual height. If similarly restrained by timber floors the effective height is equal to the actual height. Formulae are given for making allowance for restraint on vertical edges where this is known to be effective. Allowance may have to be made for the presence of openings, chases and recesses in walls.

The effective thickness of a wall of 'solid' construction is equal to the actual thickness whilst that of a cavity wall is

$$t_{ef} = (t_1^3 + t_2^3)^{1/3} \tag{4.18}$$

where t_1 and t_2 are the thicknesses of the leaves. Some qualifications of this rule are applicable if only one leaf is loaded.

The out-of-plane eccentricity of the loading on a wall is to be assessed having regard to the material properties and the principles of mechanics. A possible, simplified method for doing this is given in an Annex, but presumably any other valid method would be permissible.

An increase in the design load resistance of an unreinforced wall subjected to concentrated loading may be allowed. For walls built with units having a limited degree of perforation, the maximum design compressive stress in the locality of a beam bearing should not exceed

$$(f_k/\gamma_m)\,[(1 + 0.15\,x)\,(1.5 - 1.1\,A_b/A_{ef})] \tag{4.19}$$

where $x = 2a_1/H \not> 1$, H, A_b and A_{ef} are as shown in Fig. 4.7.

This value should be greater than the design strength f_k/γ_m but not greater than 1.25 times the design strength when $x = 0$ or 1.5 times this value when $x = 1.5$. No increase is permitted in the case of masonry built with perforated units or in shell-bedded masonry.

(d) Design of shear walls

Rather lengthy provisions are set out regarding the conditions which may be assumed in the calculation of the resistance of shear walls but the essential requirement is that the design value of the applied shear load, V_{Sd}, must not exceed the design shear resistance, V_{Rd}, i.e.

$$V_{Sd} \leqslant V_{Rd} = f_{vk} t l_c/\gamma_m \tag{4.20}$$

where f_{vk} is the characteristic shear strength of the masonry, t is the thickness of the masonry and l_c is the compressed length of the wall (ignoring any part in tension).

Distribution of shear forces amongst interconnected walls may be by elastic analysis and it would appear that the effect of contiguous floor

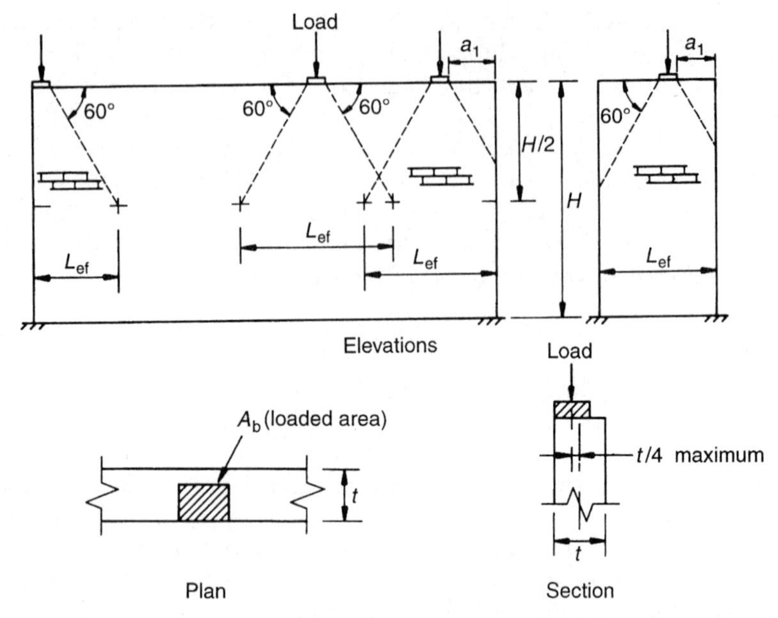

Fig. 4.7 Walls subjected to concentrated load.

slabs and intersecting walls can be included provided that the connection between these elements and the shear wall can be assured.

(e) Walls subject to lateral loading

In accordance with general principles, a wall subjected to lateral load under the ultimate limit state must have a design strength not less than the design lateral load effect. Approximate methods for ensuring this are said to be available although where thick walls are used it may not be necessary to verify the design. The provisions in ENV 1996 for lateral load design for resistance to wind loads are the same as those in BS 5628: Part 1 (1994) and need not be repeated here.

(f) Reinforced masonry

In general, the principles set out for the design of reinforced masonry follow those used for reinforced concrete and for reinforced masonry in BS 5628: Part 2, although differing slightly in detail from the latter.

The formulae for the design moment of resistance of a singly reinforced section are the same as in BS 5628 although the limit in the British code to exclude compression failures has been omitted. The provisions for shear reinforcement are, however, more elaborate and provide for the possible inclusion of diagonal reinforcement, which is uncommon in reinforced masonry sections.

A section is included on the design of reinforced masonry deep beams which may be carried out by an appropriate structural theory or by an approximate theory which is set out in some detail. In this method the lever arm, z, for calculating the design moment of resistance is, referring to Fig. 4.8, the lesser of

$$0.7l_{ef} \quad \text{or} \quad (0.4h + 0.2l_{ef}) \tag{4.21}$$

where l_{ef} is the effective span, taken to be 1.15 × the clear span, and h is the clear height of the wall.

The reinforcement A_s required in the bottom of the deep beam is then

$$A_s = M_{Rd}\,\gamma_s / f_{yk} z \tag{4.22}$$

where M_{Rd} is the design bending moment and f_{yk} is the characteristic strength of the reinforcement. The code also calls for additional nominal bed-joint reinforcement to a height of $0.5l$ above the main reinforcement or $0.5d$, whichever is the lesser, 'to resist cracking'. In this case, an upper limit of $0.4f_k bd^2/\gamma_m$ is specified although a compression failure in a deep beam seems very improbable.

Other clauses deal with serviceability and with prestressed masonry. The latter, however, refer only to ENV 1992-1-1 which is the Eurocode for prestressed concrete and give no detailed guidance.

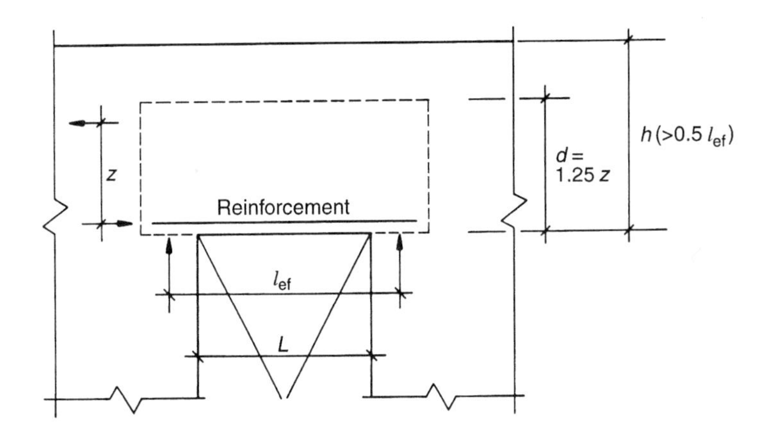

Fig. 4.8 Representation of a deep beam.

4.4.5 Sections 5 and 6: structural detailing and construction

Section 5 of ENV 1996-1-1 is concerned with detailing, making recommendations for bonding, minimum thicknesses of walls, protection of reinforcement, etc.

Section 6 states some general requirements for construction such as handling and storage of units and other materials, accuracy limits, placing of movement joints and daily construction height.

5 Design for compressive loading

5.1 INTRODUCTION

This chapter deals with the compressive strength of walls and columns which are subjected to vertical loads arising from the self-weight of the masonry and the adjacent supported floors. Other in-plane forces, such as lateral loads, which produce compression are dealt with in Chapter 6.

In practice, the design of loadbearing walls and columns reduces to the determination of the value of the characteristic compressive strength of the masonry (f_k) and the thickness of the unit required to support the design loads. Once f_k is calculated, suitable types of masonry/mortar combinations can be determined from tables, charts or equations.

As stated in Chapter 1 the basic principle of design can be expressed as

$$\text{design vertical loading} \leqslant \text{design vertical load resistance}$$

in which the term on the left-hand side is determined from the known applied loading and the term on the right is a function of f_k, the slenderness ratio and the eccentricity of loading.

5.2 WALL AND COLUMN BEHAVIOUR UNDER AXIAL LOAD

If it were possible to apply pure axial loading to walls or columns then the type of failure which would occur would be dependent on the slenderness ratio, i.e. the ratio of the effective height to the effective thickness. For short stocky columns, where the slenderness ratio is low, failure would result from compression of the material, whereas for long thin columns and higher values of slenderness ratio, failure would occur from lateral instability.

A typical failure stress curve is shown in Fig. 5.1.

The actual shape of the failure stress curve is also dependent on the properties of the material, and for brickwork, in BS 5628, it takes the form of the uppermost curve shown in Fig. 4.4 but taking the vertical

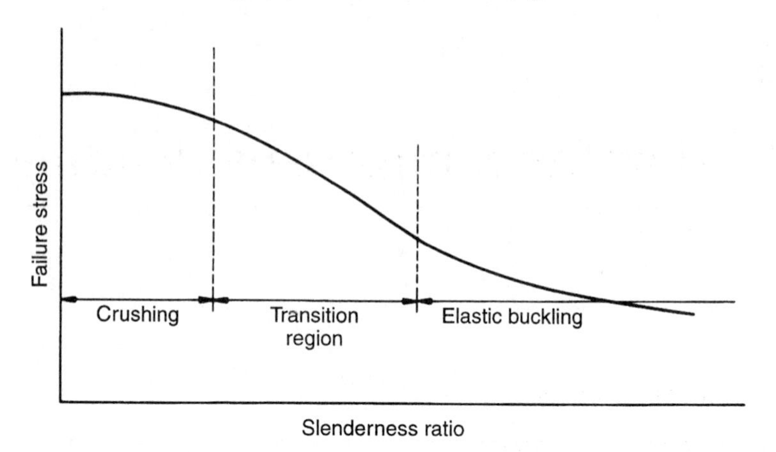

Fig. 5.1 Failure stress plotted against slenderness ratio.

axis to represent the failure stress rather than β. The failure stress at zero slenderness ratio is dependent on the strength of masonry units and mortar used in the construction and varies between 7.0 and 24 N/mm^2.

5.3 WALL AND COLUMN BEHAVIOUR UNDER ECCENTRIC LOAD

It is virtually impossible to apply an axial load to a wall or column since this would require a perfect unit with no fabrication errors. The vertical load will, in general, be eccentric to the central axis and this will produce a bending moment in the member (Fig. 5.2).

The additional moment can be allowed for in two ways:

1. The stresses due to the equivalent axial loads and bending moments can be added using the formula

$$\text{total stress} = P/A \pm M/Z$$

where A and Z are the area and section modulus of the cross-section.

2. The interaction between the bending moment and the applied load can be allowed for by reducing the axial load-carrying capacity, of the wall or column, by a suitable factor.

5.3.1 BS 5628

The second method is used in BS 5628. The effects of slenderness ratio and eccentricity are combined and appear in the code as the *capacity reduction factor* β. In the code values of β are given in tabular form based on the slenderness ratio and the eccentricity, and the equation for

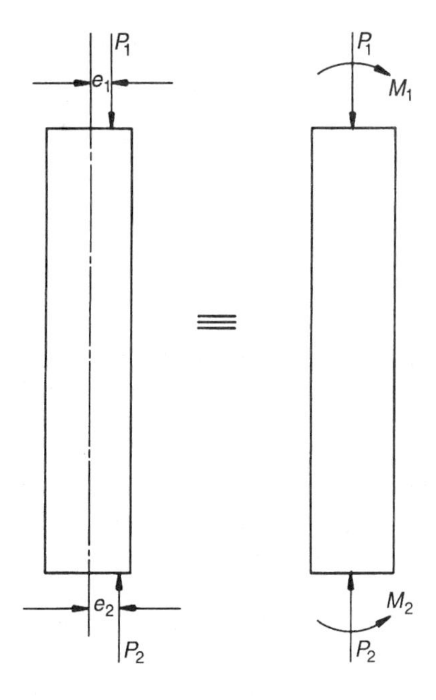

Fig. 5.2 Eccentric axial loading.

calculating the tabular values is given in Appendix B1 of the code as:

$$\beta = 1.1\,(1 - 2\,e_m/t) \tag{5.1}$$

where e_m is the larger value of e_x, the eccentricity at the top of the wall, and e_t, the eccentricity in the mid-height region of the wall. Values of e_t are given by the equation:

$$e_t = 0.6e_x + e_a = 0.6e_x + t\,[(1/2400)\,(h_{ef}/t)^2 - 0.015] \tag{5.2}$$

where (h_{ef}/t) is the slenderness ratio (section 5.4) and e_a represents an additional eccentricity to allow for the effects of slenderness.

A graph showing the variation of β with slenderness ratio and eccentricity was shown previously in Fig. 4.4 and further details of the method used for calculating β are given in sections 5.6.2 and 5.9.

5.3.2 ENV 1996-1-1

A similar approach is used in the Eurocode, ENV 1996-1-1, except that a capacity reduction factor Φ is used instead of β. The effects of slenderness and eccentricity of loading are allowed for in both Φ and β but in a slightly different way. In the Eurocode, values of Φ_i at the top (or bottom) of the wall are defined by an equation similar to that given in BS

5628 whilst values of Φ_m in the mid-height region are determined from a set of curves (Fig. 4.6).

1. At the top (or bottom) of the wall values of Φ are defined by

$$\Phi_i = 1 - 2\,(e_i/t) \tag{5.3}$$

where

$$e_i = M_i/N_i + e_{hi} + e_a \geqslant 0.05t \tag{5.4}$$

where, with reference to the top (or bottom) of the wall, M_i is the design bending moment, N_i the design vertical load, e_{hi} the eccentricity resulting from horizontal loads, e_a the accidental eccentricity and t the wall thickness. The accidental eccentricity e_a, which allows for construction imperfections, is assumed to be $h_{ef}/450$ where h_{ef} is the effective height. The value 450, representing an average 'category of execution', can be changed to reflect a value more appropriate to a particular country.

2. For the middle fifth of the wall Φ_m can be determined from Fig. 4.6 using values of h_{ef}/t_{ef} and e_{mk}/t. Figure 4.6, used in EC6, is equivalent to Fig. 4.4, used in BS 5628, to obtain values of Φ and β respectively. The value of e_{mk} is obtained from:

$$e_{mk} = M_m/N_m + e_{hm} \pm e_a + e_k \geqslant 0.005t \tag{5.5}$$

where, with reference to the middle one-fifth of the wall height, M_m is the design bending moment, N_m the design vertical load, e_{hm} the eccentricity resulting from horizontal loads and e_k the creep eccentricity defined by

$$e_k = 0.002\Phi_\infty\,(h_{ef}/t_m)\,(te_m)^{1/2}$$

where Φ_∞ is a final creep coefficient obtained from a table given in the code. However, the value of e_k can be taken as zero for all walls built with clay and natural stone units and for walls having a slenderness ratio up to 15 constructed from other masonry units.

Note that the notation e_a used in EC6 is not the same quantity e_a used in BS 5628. They are defined and calculated differently in the two codes.

5.4 SLENDERNESS RATIO

This is the ratio of the effective height to the effective thickness, and therefore both of these quantities must be determined for design purposes. The maximum slenderness ratio permitted according to both BS 5628 and ENV 1996-1-1 is 27.

5.4.1 Effective height

The effective height is related to the degree of restraint imposed by the floors and beams which frame into the wall or columns.

Theoretically, if the ends of a strut are free, pinned, or fully fixed then, since the degree of restraint is known, the effective height can be calculated (Fig. 5.3) using the Euler buckling theory.

In practice the end supports to walls and columns do not fit into these neat categories, and engineers have to modify the above theoretical values in the light of experience. For example, a wall with concrete floors framing into the top and bottom, from both sides (Fig. 5.4), could be considered as partially fixed at both ends, and for this case the effective length is taken as $0.75h$, i.e. half-way between the 'pinned both ends' and the 'fixed both ends' cases.

In the above example it is assumed that the degree of fixity is half-way between the pinned and fixed case, but in reality the degree of fixity is dependent on the relative values of the stiffnesses of the floors and walls. For the case of a column with floors framing into both ends, the stiffnesses of the floors and columns are of a similar magnitude and the effective height is taken as h, the clear distance between lateral supports (Fig. 5.4).

(a) BS 5628

In BS 5628 the effective height is related to the degree of lateral resistance to movement provided by supports, and the code distinguishes between two types of resistance – simple and enhanced. The term *enhanced resistance* is intended to imply that there is some degree of rotational restraint at the end of the member. Such resistances would arise, for example, if floors span to a wall or column from both sides at the same level or where a concrete floor on one side only has a bearing greater than 90 mm and the building is not more than three storeys.

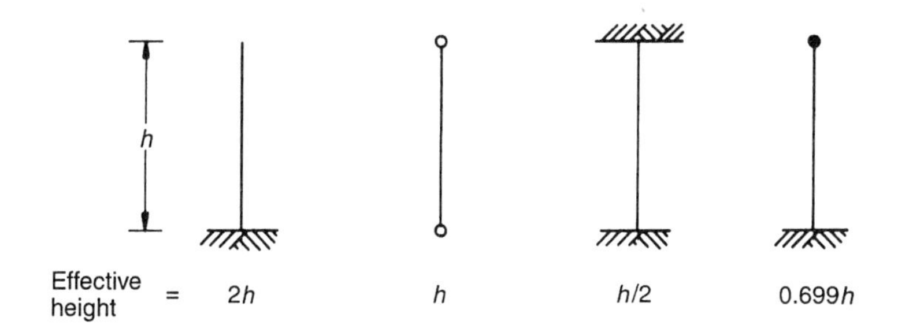

Fig. 5.3 Effective height for different end conditions.

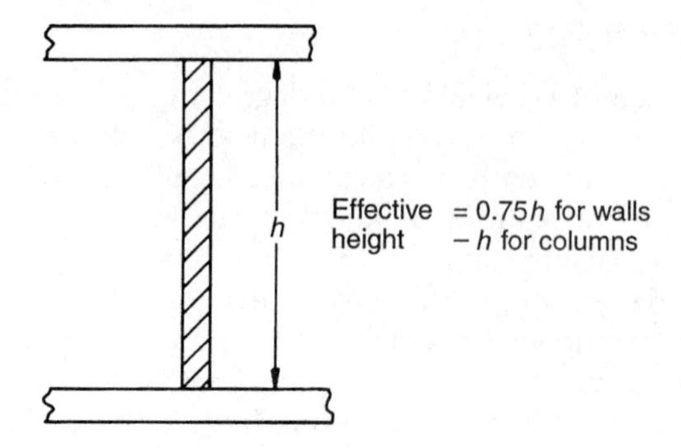

Fig. 5.4 Effective height for wall/floor and wall/column arrangement.

Conventional values of effective height recommended in BS 5628 are:

- *Walls*
  Enhanced resistance

$$h_{ef} = 0.75h$$

  Simple resistance

$$h_{ef} = h$$

- *Columns*
  With lateral supports in two directions

$$h_{ef} = h$$

  With lateral support in one direction

$$h_{ef} = h \text{ (in lateral support direction)}$$

$$h_{ef} = 2h \text{ (in direction in which support is not provided)}$$

- *Columns formed by adjacent openings in walls*
  Enhanced resistance

$$h_{ef} = 0.75h + 0.25 \times \text{(height of the taller of the two openings)}$$

  Simple resistance

$$h_{ef} = h$$

(b) ENV 1996-1-1

In the Eurocode the effective height is taken as:

$$h_{ef} = \rho_n h \qquad (5.6)$$

where h is the clear storey height and ρ_n is a reduction factor where $n=2, 3$ or 4 depending on the edge restraint or stiffening of the wall. Suggested values of ρ_n given in the code are:

- For walls restrained at the top and bottom then

$$\rho_2 = 0.75 \text{ or } 1.0 \quad \text{depending on the degree of restraint}$$

- For walls restrained top and bottom and stiffened on one vertical edge with the other vertical edge free

$$\rho_3 = \rho_2/[1+(\rho_2 h/3L)^2] > 0.3 \quad \text{when } h \leqslant 3.5L$$

$$\rho_3 = 1.5L/h \quad \text{when } h > 3.5L$$

  where L is the distance of the free edge from the centre of the stiffening wall. If $L \geqslant 15t$, where t is the thickness of the stiffened wall, take $\rho_3 = \rho_2$.

- For walls restrained top and bottom and stiffened on two vertical edges

$$\rho_4 = \rho_2/[1+(\rho_2 h/L)^2] \quad \text{when } h \leqslant L$$

$$\rho_4 = 0.5L/h \quad \text{when } h > L$$

  where L is the distance between the centres of the stiffening walls. If $L \geqslant 30t$, where t is the thickness of the stiffened wall, take $\rho_4 = \rho_2$.

Note that walls may be considered as stiffened if cracking between the wall and the stiffening is not expected or if the connection is designed to resist developed tension and compression forces by the provision of anchors or ties. These conditions are important and designers should ensure that they are satisfied before assuming that any stiffening exists. Stiffening walls should have a length of at least one-fifth of the storey height and a thickness of 0.3 × (wall thickness) with a minimum value of 85 mm.

5.4.2 Effective thickness

The effective thickness of single leaf walls or columns is usually taken as the actual thickness, but for cavity walls or walls with piers other assumptions are made.

(a) BS 5628

Considering the single leaf wall with piers shown in Fig. 5.5(a) it is necessary to decide on the value of the factor K shown in Fig. 5.5(b), which will give a wall of equivalent thickness. Here, the meaning of

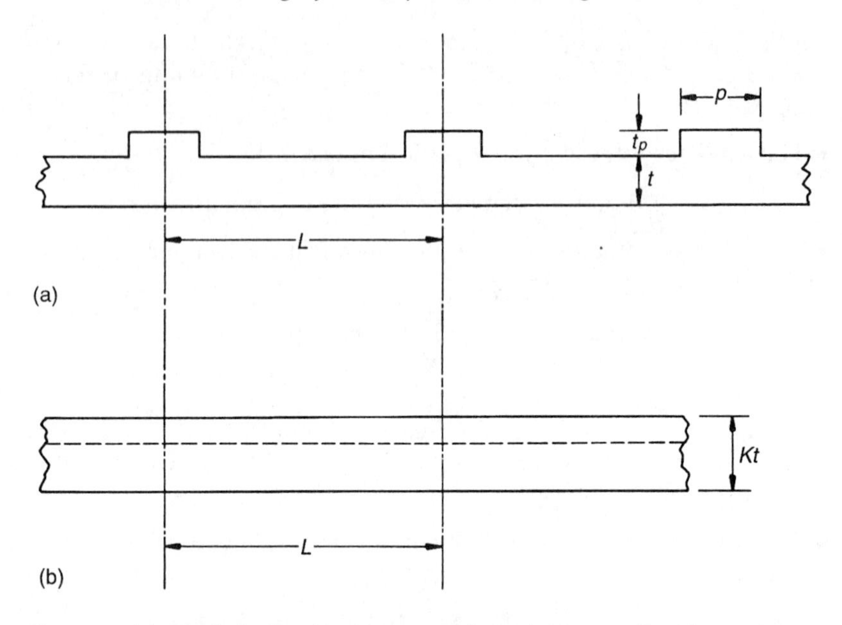

Fig. 5.5 (a) Single leaf wall with piers; (b) equivalent wall without piers.

'equivalent' is vague since it implies some unknown relationship between the areas and section moduli for the two cases.

Suggested values for K are given in BS 5628, and these are reproduced below in Table 5.1.

The effective thickness for cavity walls is taken as the greater value of two-thirds the sum of the actual thicknesses of the two leaves or the actual thickness of the thicker leaf. For the case of a cavity wall with piers a similar calculation, but introducing the factor K from Table 5.1, is used (Fig. 5.6).

Table 5.1 K values for effective thickness of walls stiffened by piers

	t_p/t		
L/p	1	2	3
6	1.0	1.4	2.0
10	1.0	1.2	1.4
20	1.0	1.0	1.0

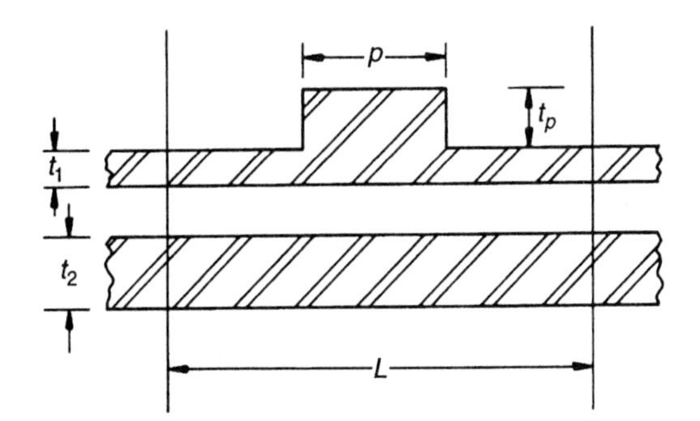

Fig. 5.6 Cavity wall with piers.

Effective thickness is taken as the greatest value of:

- $\frac{2}{3}(t_1 + Kt_2)$
- t_1
- Kt_2

According to the code the stiffness coefficients given in Table 5.1 can also be used for a wall stiffened by intersecting walls if the assumption is made that the intersecting walls are equivalent to piers of width equal to the thickness of the intersecting walls and of thickness equal to three times the thickness of the stiffened wall. However, recent experiments do not confirm this. A series of tests conducted by Sinha and Hendry on brick walls stiffened either by returns or by intersecting diaphragm walls under axial compressive loading showed no increase in strength compared to strip walls for a range of slenderness ratios up to 32.

(b) ENV 1996-1-1

In the Eurocode the effective thickness of a cavity wall in which the leaves are connected by suitable wall ties is determined using:

$$t_{ef} = (t_1^3 + t_2^3)^{1/3} \tag{5.7}$$

5.5 CALCULATION OF ECCENTRICITY

In order to determine the value of the eccentricity, different simplifying assumptions can be made, and these lead to different methods of calculation. The simplest is the approximate method given in BS 5628, but a more accurate value can be obtained, at the expense of additional calculation, by using a frame analysis. Calculation of the eccentricity according to the Eurocode is performed using the equations given in section 5.3. The approach using these equations is similar to the method given in BS 5628.

5.5.1 Approximate method of BS 5628

1. The load transmitted by a single floor is assumed to act at one-third of the depth of the bearing areas from the face of the wall (Figs. 5.7(a) and (b)).
2. For a continuous floor, the load from each side is assumed to act at one-sixth of the thickness of the appropriate face (Fig. 5.8 (a)).
3. Where joist hangers are used the load is assumed to act at the centre of the joist bearing areas of the hanger (Fig. 5.8(b)).
4. If the applied vertical load acts between the centroid of the two leaves of a cavity wall it should be replaced by statically equivalent axial loads in the two leaves (Fig. 5.9).

In the above the total vertical load on a wall, above the lateral support being considered, is assumed to be axial.

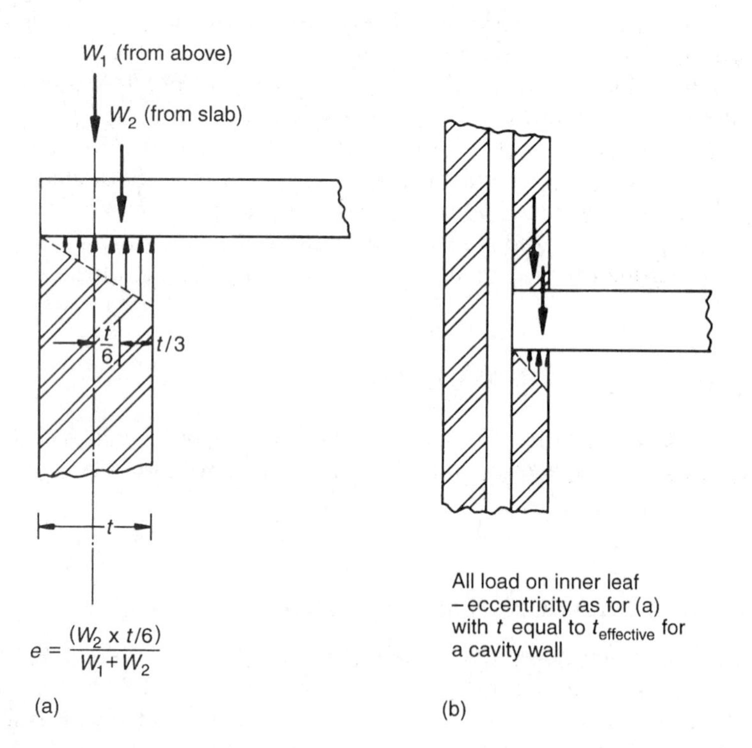

Fig. 5.7 (a) Eccentricity for floor/solid wall; (b) eccentricity for floor/cavity wall.

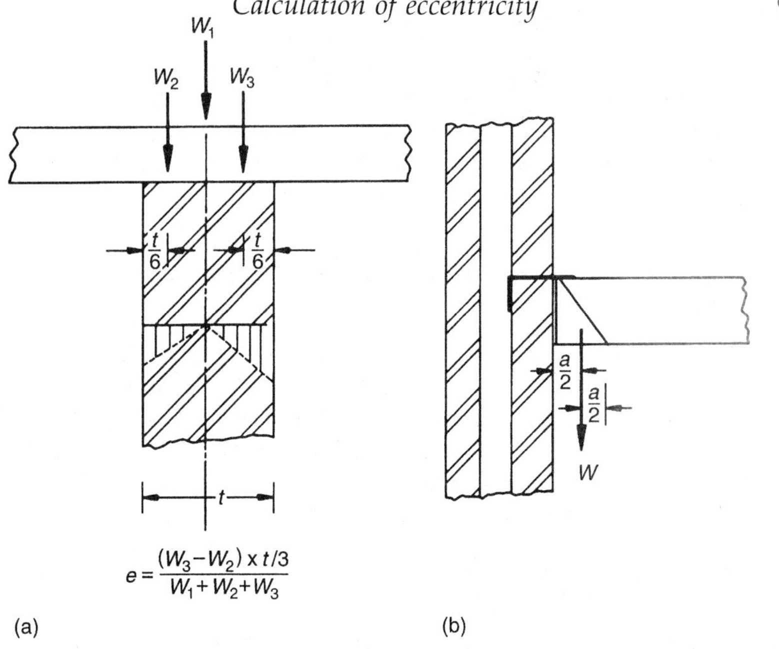

Fig. 5.8 (a) Eccentricity for continuous floor/wall; (b) assumed load position with joist hanger.

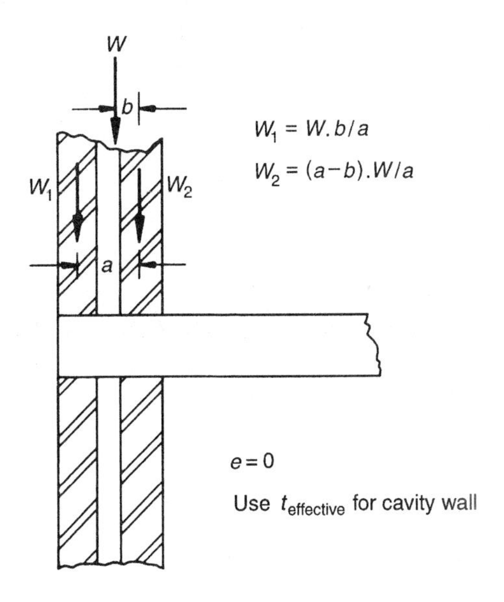

Fig. 5.9 Eccentricity for cavity wall.

Note that the eccentricity calculated above is the value at the top of the wall or column where the floor frames into the wall. In BS 5628 the eccentricity is assumed to vary from the calculated value at the top of the wall to zero at the bottom of the wall, subject to an additional eccentricity being considered to cover slenderness effects (see Chapter 4).

5.5.2 Simplified method for calculating the eccentricity (ENV 1996-1-1)

In order to calculate the eccentricities e_i or e_m it is necessary to determine the value of M_i or M_m and a simplified method of calculating these moments is described in Annex C of EC6. Using the simplified frame diagram illustrated in Fig. 5.10 in which the remote ends of each member framing into a joint are assumed to be fixed (unless known to be free), the bending moment M_1 can be calculated using:

$$M_1 = \frac{nE_1I_1/h_1}{nE_1I_1/h_1 + nE_2I_2/h_2 + nE_3I_3/L_3 + nE_4I_4/L_4}\left(\frac{w_3L_3^2}{12} - \frac{w_4L_4^2}{12}\right) \quad (5.8)$$

where n is taken as 4 if the remote end is fixed and 3 if free. The value of M_2 can be obtained from the same equation but replacing the numerator with nE_2I_2/h_2. Here E and I represent the appropriate modulus of elasticity and second moment of area respectively, and w_3 and w_4 are the design uniformly distributed loads modified by the partial safety factors. If less than four members frame into a joint then the equation is modified by ignoring the terms related to the missing members.

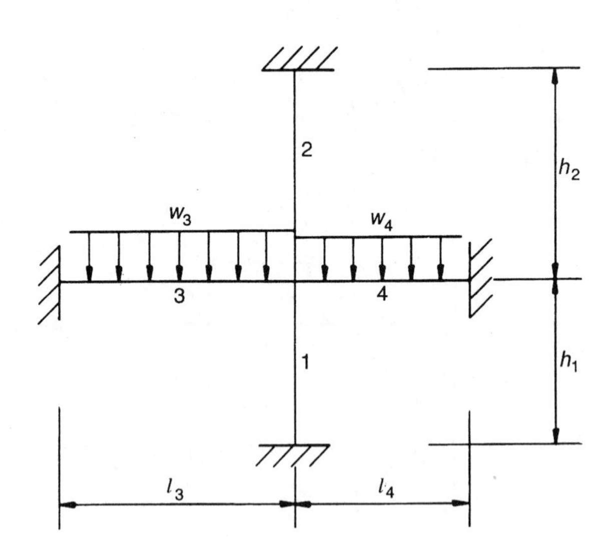

Fig. 5.10 Simplified frame diagram.

The code states that this simplified method is not suitable for timber floor joists and proposes that for this case the eccentricity be taken as $\geqslant 0.4t$. Also, since the results obtained from the equation tend to be conservative the code allows the use of a reduction factor $(1-k/4)$ if the design vertical stress is greater than $0.25\ \mathrm{N/mm^2}$. The value of k is given by

$$k = (k_3 + k_4)/(k_1 + k_2) \tag{5.9}$$

where each k is the stiffness factor defined by EI/h.

5.5.3 Frame analysis

If the wall bending moment and axial load are calculated for any joint in a multi-storey framed structure then the eccentricity can be determined by dividing the moment by the axial load. The required moment and axial load can be determined using a normal rigid frame analysis. This approach is reasonable when the wall compression is high enough to contribute to the rigidity of the joints, but would lead to inaccuracies when the compression is small.

The complete frame analysis can be avoided by a partial analysis which assumes that the far ends of members (floors and walls) attached to the joint under consideration are pinned (Fig. 5.11).

The wall bending moments for the most unfavourable loading conditions can now be determined using moment-distribution or slope-deflection methods.

More sophisticated methods which allow for the relative rotation of the wall and slab at the joints and changing wall stiffness due to tension cracking in flexure are being developed.

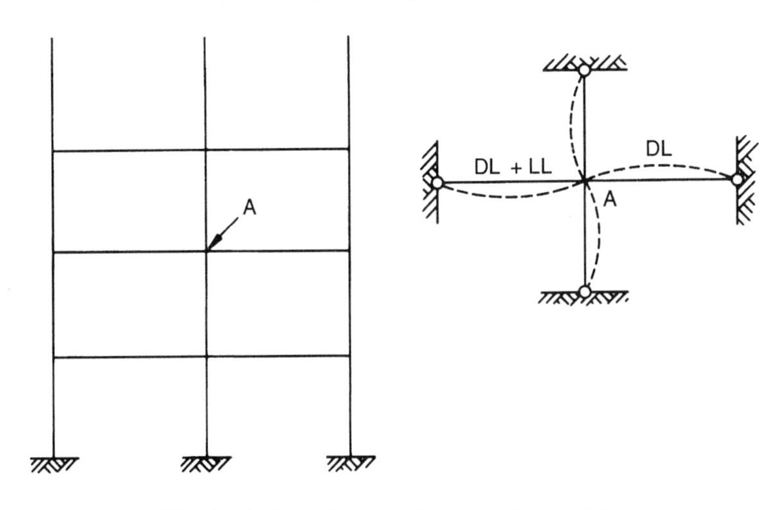

Fig. 5.11 Multi-storey frame and typical joint.

5.6 VERTICAL LOAD RESISTANCE

The resistance of walls or columns to vertical loading is obviously related to the characteristic strength of the material used for construction, and it has been shown above that the value of the characteristic strength used must be reduced to allow for the slenderness ratio and the eccentricity of loading. If we require the *design* vertical load resistance, then the characteristic strength, which is related to the strength at failure, must be further reduced by dividing by a safety factor for the material.

As shown in Chapter 4 the British code introduces a capacity reduction factor β which allows simultaneously for effects of eccentricity and slenderness ratio. It should be noted that these values of β are for use with the assumed notional values of eccentricity given in the code, and that if the eccentricity is determined by a frame type analysis which takes account of continuity then different capacity reduction factors should be used.

As shown in section 5.3 the Eurocode introduces the capacity reduction factor Φ which is similar to, but not identical with, the factor β used in BS 5628.

If tensile strains are developed over part of a wall or column then there is a reduction in the effective area of the cross-section since it can be assumed that the area under tension has cracked. This effect is of importance for high values of eccentricity and slenderness ratio, and the Swedish code allows for it by introducing the ultimate strain value for the determination of the reduction factor.

5.6.1 Design vertical load resistance of walls

Using the principles outlined above the design vertical load resistance per unit length of wall is given in BS 5628 as $(\beta t f_k)/\gamma_m$ where γ_m is the partial safety factor for the material and β is obtained from Fig. 4.4. In the Eurocode the design vertical load resistance per unit length of wall is given as $(\Phi t f_k)/\gamma_m$ where Φ is determined either at the top (or bottom), Φ_i, or in the middle fifth of the wall, Φ_m.

The procedure for calculating the design vertical load resistance in BS 5628 can be summarized as follows:

1. Determine e_x at the top of the wall using the method illustrated in Figs 5.7 to 5.9.
2. Determine e_a, the additional eccentricity, using equation (4.2) and the total eccentricity e_t using equation (4.3).
3. If $e_x > e_t$ then e_x governs the design. If $e_t > e_x$ then e_t (the eccentricity at mid-height) governs.
4. Taking e_m to represent the larger value of e_x and e_t, then if e_m is $\leqslant 0.05t$ the design load resistance is given by $(\beta t f_k)/\gamma_m$, with $\beta = 1$, and if e_m

$> 0.05t$ the design load resistance is given by $(\beta t f_k)/\gamma_m$, with $\beta = 1.1\,(1 - 2e_m/t)$.

5.6.2 Design vertical load resistance of columns

For columns the design vertical load resistance is given in BS 5628 as $(\beta t f_k)/\gamma_m$, but for this case the rules in Table 5.2 apply to the selection of β from Fig. 4.4.

If the eccentricities at the top of the column about the major and minor axes are greater than $0.05b$ and $0.05t$ respectively, then the code recommends that the values of β can be determined from the equations given in Appendix B of BS 5628. The method can be summarized as follows (Fig. 5.12):

1. About XX axis

- The design eccentricity e_m about XX is defined as the larger value of e_x and e_t, where

$$e_t = 0.6\,e_x + t\,[(1/2400)\,(h_{ef}/t)^2 - 0.015]$$

and (h_{ef}/t) is the slenderness ratio about the minor axis.

- The value of β is calculated from

$$\beta = 1.1\,(1 - 2e_m/t)$$

Table 5.2 Rules for selecting β for columns

Eccentricity at top of column about major axis	*Eccentricity at top of column about minor axis*	*Selection of* β
$<0.05b$	$<0.05t$	Use upper curve of Fig. 4.4 with t_{ef} appropriate to minor axis
$<0.05b$	$>0.05t$	Use Fig. 4.4 with both eccentricity and slenderness ratio appropriate to minor axis
$>0.05b$	$<0.05t$	Use Fig. 4.4 with eccentricity appropriate to major and slenderness ratio to minor axis
$>0.05b$	$>0.05t$	See text

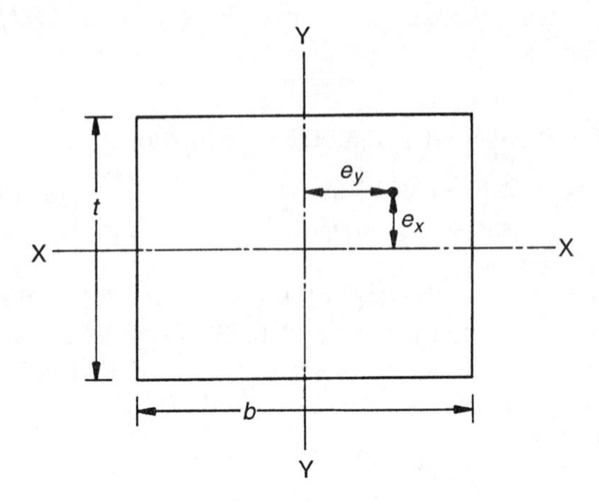

Fig. 5.12 Column cross-section.

2. About YY axis

- Use a similar procedure using e_y and the slenderness ratio about the major axis. Note that no slenderness effect need be considered when the slenderness ratio is less than 6 (see example below).

Example
Determine the values of β for a solid brickwork column of cross-section 215 mm × 430 mm (Fig. 5.13) and effective height about both axes of 2500 mm if the eccentricities at the top of the columns about the major and minor axes are (a) 25 mm and 10 mm and (b) 60 mm and 20 mm respectively.

Solution (a)

$$e_x = 10 = 0.046t \quad \text{i.e.} < 0.05t$$

$$e_y = 25 = 0.058b \quad \text{i.e.} > 0.05b$$

Therefore use Fig. 4.4 with eccentricity appropriate to major axis YY (25 mm) and slenderness ratio appropriate to minor axis. Slenderness ratio $SR = 2500/215 = 11.63$. Using Fig. 4.4, $\beta \approx 0.93$.

Solution (b)

$$e_x = 20 = 0.093t \quad \text{i.e.} > 0.05t$$

$$e_y = 60 = 0.139b \quad \text{i.e.} > 0.05b$$

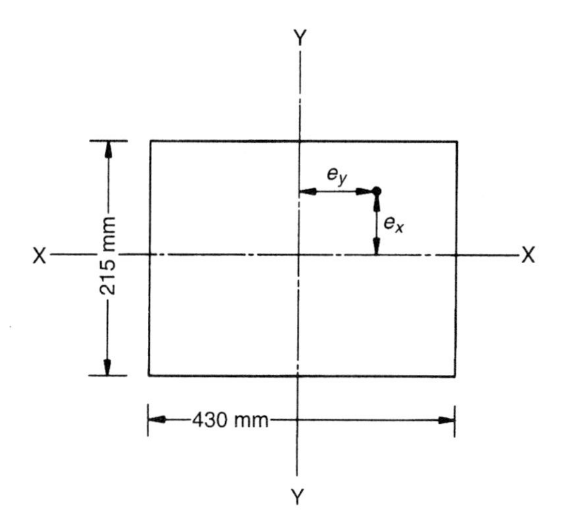

Fig. 5.13 Dimensions of worked example.

About XX axis

$$e_m = e_x = 20 \text{ mm}$$

or

$$e_m = e_t = 0.6 \times 20 + 215\,[\,(1/2400)(2500/215)^2 - 0.015\,]$$
$$= 12 + 8.89 = 20.89 \text{ mm}$$

So

$$\beta_{XX} = 1.1\,[1 - (2 \times 20.89/215)\,] = 0.89$$

About YY axis

$$e_m = e_y = 60 \text{ mm}$$

or

$$e_m = e_t = 0.6 \times 60 + 430\,[(1/2400) \times (2500/430)^2 - 0.015\,]$$
$$= 36 + 430\,(0.014 - 0.015)$$

For this case the bracketed term is negative, because the slenderness ratio is less than 6, and therefore no additional term due to slenderness effect is required. That is $e_m = 60$ mm and

$$\beta_{YY} = 1.1\,[1 - (12 \times 60/430)] = 0.79$$

Note that the design vertical load resistance for the above example would be

(a) $(\beta t f_k)/\gamma_m = 0.93 \times 215 \times 430 \times f_k/\gamma_m$

(b) $(\beta t f_k)/\gamma_m = 0.89 \times 215 \times 430 \times f_k/\gamma_m$

That is, the largest value of β_{XX} and β_{YY} is used in order to ensure that the smaller value of f_k will be determined when the design vertical load resistance is equated to the design vertical load.

No specific references to the design of columns are given in the Eurocode although a similar approach to that outlined above but replacing β with Φ would be possible.

5.6.3 Design vertical load resistance of cavity walls or columns

The design vertical load resistance for cavity walls or columns can be determined using the methods outlined in sections 5.6.1 and 5.6.2 if the vertical loading is first replaced by the statically equivalent axial load on each leaf. The effective thickness of the cavity wall or column is used for determining the slenderness ratio for each leaf of the cavity.

5.6.4 Design vertical strength for concentrated loads

Increased stresses occur beneath concentrated loads from beams and lintels, etc. (see Fig. 4.5), and the combined effect of these local stresses with the stresses due to other loads should be checked. The concentrated load is assumed to be uniformly distributed over the bearing area.

(a) BS 5268

In BS 5628 two design checks are suggested:

- At the bearing, assuming a local design bearing strength of either $1.25 f_k/\gamma_m$ or $1.5 f_k/\gamma_m$ depending on the type of bearing.
- At a distance of $0.4h$ below the bearing, where the design strength is assumed to be $\beta f_k/\gamma_m$. The concentrated load is assumed to be dispersed within a zone contained by lines extending downwards at 45° from the edges of the loaded area (Fig. 5.14).

The code also makes reference to the special case of a spreader beam located at the end of a wall and spanning in its plane. For this case the maximum stress at the bearing, combined with stresses due to other loads, should not exceed $2.0 f_k/\gamma_m$.

(b) ENV 1996-1-1

In ENV 1996-1-1 the following checks are suggested:

- For Group 1 masonry units, the local design bearing strength must not exceed the value derived from

$$(f_k/\gamma_m)\{(1+0.15x)[1.5-1.1(A_b/A_{ef})]\} \tag{5.10}$$

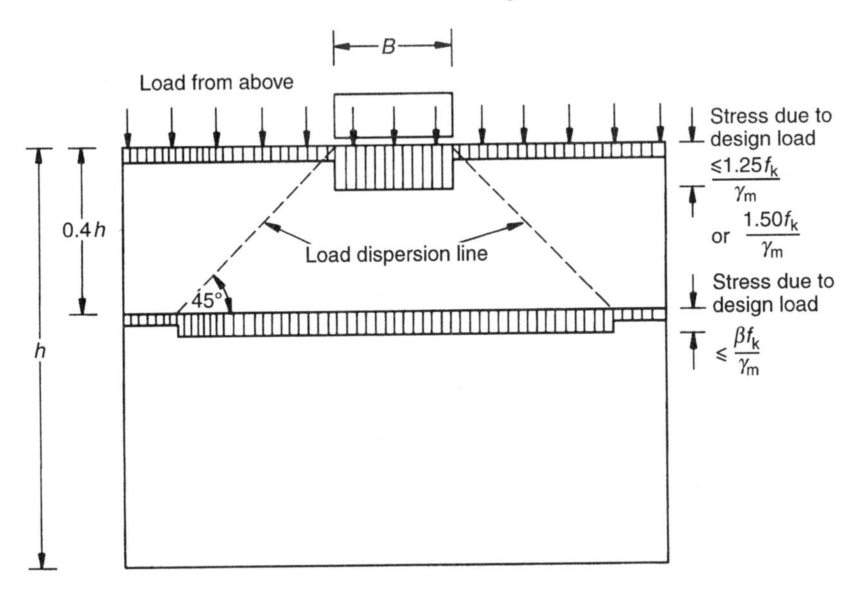

Fig. 5.14 Stress distribution due to concentrated load (BS 5628).

but not less than f_k/γ_m nor greater than either $1.25f_k/\gamma_m$ or $1.5f_k/\gamma_m$ depending on the type of bearing. Here $x = 2a_1/H$ but $x < 1.0$, a_1 is the distance from the end of the wall to the nearer edge of the bearing, H is the height of the wall to the level of the load, A_b is the bearing area of load but $A_b < 0.45A_{ef}$ and A_{ef} is the effective area of wall $L_{ef}t$.

For Groups 2a, 2b and 3 masonry units, the design strength should not exceed f_k/γ_m.

- At a distance of $0.5H$ below the bearing the design strength is assumed to be $\Phi f_k/\gamma_m$ and the requirements of section 5.6.1 should be met. The concentrated load is assumed to be dispersed within a zone contained by lines extending downwards at 60° from the edges of the loaded area.

The code also makes reference to the special case of a spreader beam of width t, height greater than 200 mm and length greater than three times the bearing length of the load. For this case the maximum stress beneath the loaded area should not exceed $1.5f_k/\gamma_m$.

5.7 VERTICAL LOADING

Details about characteristic dead and imposed loads and partial safety factors have been given in Chapter 4, and values of the design vertical loads will already have been determined for the calculation of the eccentricities. The design process for vertical loading is completed by equating the design vertical loading to the appropriate design vertical load resistance and using the resulting equation to determine the value of the characteristic compressive strength of the masonry f_k. Typically the equation takes the form

$$\Sigma W \text{ kN/m} = (\beta t f_k)/\gamma_m \tag{5.11}$$

Generally the calculation of ΣW involves the summation of products of the partial safety factor for load (γ_f) with the appropriate characteristic load (G_k and Q_k). This is discussed in Chapter 4 and illustrated in Chapter 10. For design according to the Eurocode, β in equation (5.11) would be replaced by Φ.

Using standard tables or charts and modification factors where applicable, the compressive strength of the masonry units and the required mortar strength to provide the necessary value of f_k can be obtained.

Examples of the calculation for an inner solid brick wall and an external cavity wall are given in section 5.9.

5.8 MODIFICATION FACTORS

The value of f_k used in Fig. 4.1, in order to determine a suitable masonry/mortar combination, is sometimes modified to allow for the effects of small plan area or narrow masonry walls.

5.8.1 Small plan area

(a) BS 5628

If the horizontal cross-sectional area (A) is less than 0.2 m^2 then the value of f_k determined from an equation similar to (5.11) is divided by a factor $(0.70 + 1.5A)$.

(b) ENV 1996-1-1

If the horizontal cross-sectional area (A) is less than 0.1 m^2 then the value of f_k determined from an equation similar to (5.11) is divided by a factor $(0.70 + 3A)$.

5.8.2 Narrow masonry walls

In BS 5628 a modification factor is also given for narrow walls. If the thickness of the wall is equal to the width of the masonry then the value of f_k determined from an equation similar to (5.11) is divided by 1.15.

Note that some designers include the above modification factors in the basic equation (5.11) where they appear as a multiplication factor on the right-hand side, e.g. for narrow walls, equation (5.11) could be rewritten

$$\Sigma W = \beta t\,(1.15 f_k)/\gamma$$

5.9 EXAMPLES

5.9.1 Example 1: Internal masonry wall (Fig. 5.15)

(a) Using BS 5628

Loading (per metre run of wall)

	Dead load (kN/m)	*Imposed load* (kN/m)
Load from above	105.0	19.0
Self-weight of wall	17.0	–
Load from left slab	4.1	2.2
Load from right slab	4.1	2.2

Safety factors
For material strength, $\gamma_m = 3.5$
For loading, $\gamma_f\,(\mathrm{DL}) = 1.4$
$\gamma_f\,(\mathrm{LL}) = 1.6$

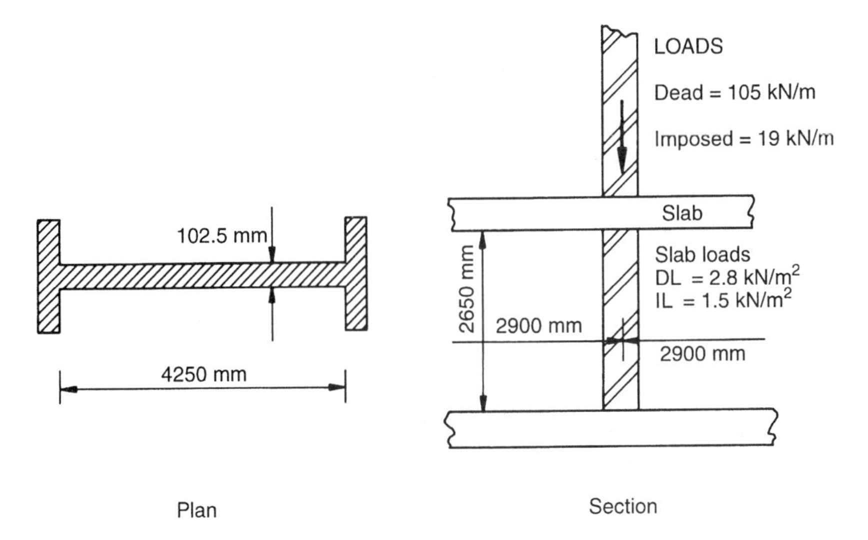

Fig. 5.15 Plan and section details for example 1.

Design vertical loading (Fig. 5.16)
Loading from above (W_1) = 1.4 × 105 + 1.6 × 19 = 177.4 kN/m
Load from left (W_2)

dead load only = 1.4 × 4.1 = 5.7 kN/m

imposed load = 5.7 + 1.6 × 2.2 = 9.2 kN/m

Load from right (W_3)

dead load only = 1.4 × 4.1 = 5.7 kN/m

imposed load = 5.7 + 1.6 × 2.2 = 9.2 kN/m

Wall self-weight = 1.4 × 17 = 23.8 kN/m

Slenderness ratio

Effective height = 0.75 × 2650 = 1988 mm

Effective thickness = actual thickness = 102.5 mm

Slenderness ratio = 1988/102.5 = 19.4

Eccentricity

See section 5.5.1

- With full DL + IL on each slab there will be no eccentricity since $W_2 = W_3$.

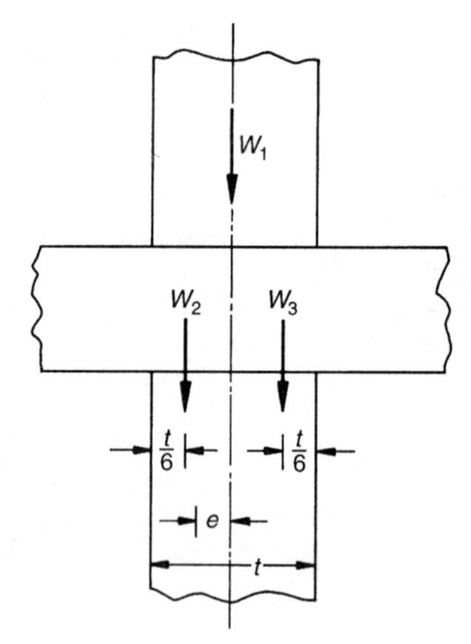

Fig. 5.16 Loading arrangement for eccentricity calculation.

- With only one slab loaded with superimposed load, $W_2 = 9.2$ and $W_3 = 5.7$.

Taking moments about centre line

$$\Sigma W e_x = W_2 (t/3) - W_3 (t/3)$$

$$e_x = (9.2 - 5.7) \times (t/3)/[9.2 + 5.7 + 177.4 + (1.4 \times 17)]$$

$$= 0.5534 \text{ mm}$$

From equation (5.2)

$$e_t = 0.6 \times 0.5534 + 102.5\,[(1/2400) \times (19.4)^2 - 0.015]$$

$$= 0.332 + 14.536 = 14.868 \text{ mm}$$

So that, since e_t is greater than e_x, $e_m = e_t = 0.145t$, which is greater than $0.05t$, with the result that:

$$\beta = 1.1\,[1 - (2 \times 0.145)] = 0.78$$

Design vertical load resistance
Assume t in mm and f_k in N/mm^2:

$$\text{design vertical load resistance} = (\beta t f_k)/\gamma_m$$

$$= 0.78 \times 102.5 \times f_k/3.5$$

$$= 22.84\, f_k \quad (\text{N/mm or kN/m})$$

Determination of f_k
We have

$$\text{design vertical load} = \text{design vertical load resistance}$$

$$(177.4 + 9.2 + 9.2 + 23.8) \text{ kN/m} = 22.84\, f_k \text{ kN/m}$$

$$f_k = 9.61 \text{ N/mm}^2$$

Modification factors for f_k

- Horizontal cross-sectional area of wall $= 0.1025 \times 4.25 = 0.44$ m^2. Since $A > 0.2$ m^2, no modification factor for area.
- Narrow masonry wall. Since wall is one brick thick, modification factor $= 1.15$.

Required value of f_k

$$f_k = 9.61/1.15 = 8.35 \text{N/mm}^2$$

Selection of brick/mortar combination

Use Fig. 4.1 to select a suitable brick/mortar combination. Any of the following would provide the required value of f_k.

Mortar designation	*Compressive strength of bricks* (N/mm²)
(iii)	35.0
(ii)	29.0
(i)	22.5

(b) Using ENV 1996-1-1

The dimensions, loadings and safety factors used here are the same as those given above in section (a). The reinforced concrete floor slabs are assumed to be of the same thickness as the walls (102.5 mm) and the modular ratio E_{slab} / E_{wall} taken as 2.

Loading
As for section (a).

Safety factors
For material strength, $\gamma_m = 3.0$
For loading, γ_f (DL) = 1.35
γ_f (LL) = 1.5

Design vertical loading (Fig. 5.16)

Loading from above (W_1) = 1.35 × 105 + 1.5 × 19 = 170.25 kN/m

Load from left (W_2)

$$\text{dead load only} = 1.35 \times 4.1 = 5.535 \text{ kN/m}$$

$$\text{imposed load} = 5.535 + 1.5 \times 2.2 = 8.835 \text{ kN/m}$$

Load from right (W_3)

$$\text{dead load only} = 5.535 \text{ kN/m}$$

$$\text{imposed load} = 8.835 \text{ kN/m}$$

$$\text{Wall self-weight} = 1.35 \times 17 = 22.95 \text{ kN/m}$$

Eccentricity
Because of the symmetry equation (5.8) can be rewritten:

$$M_1 = \frac{4\,E_w I_w/h}{8\,E_w I_w/h + 8\,E_c I_c/L_3}\left(\frac{w_3 L_3^2}{12} - \frac{w_2 L_2^2}{12}\right)$$

or

$$M_1 = \frac{1}{2 + 2(E_c I_c h/E_w I_w L_3)}\left(\frac{L_3^2}{12}(w_3 - w_2)\right)$$

Taking $E_c/E_w = 2$, $I_c/I_w = 1$, $h = 2650$ mm and the clear span $L_3 = 2797.5$ mm

$$M_1 = \frac{1[2.8^2 \times (8.835 - 5.535)/12]}{2 + (2 \times 2 \times 1 \times 0.947)} = \frac{2.156}{5.788} = 0.372\,\text{kN m}$$

$$N_1 = 8.835 + 5.535 + 170.25 + (1.35 \times 17) = 207.57\,\text{kN}$$

$$M_1/N_1 = 0.372/207.57 = 0.0018\,\text{m}$$

Taking $e_{hi} = 0$ and $e_a = h_{ef}/450 = 1.988/450 = 0.004\,\text{m}$ equation (5.4) becomes

$$e_i = 0.0018 + 0 + 0.004 = 0.0058 \quad (\geqslant 0.05t = 0.005)$$

The design vertical stress at the junction is 207.57/102.5 and since this is greater than $0.25\,\text{N/mm}^2$ the code allows the eccentricity to be reduced by $(1 - k/4)$ where k is given by equation (5.9).

For this example

$$k = E_c I_c h/E_w I_w L_3 = 2 \times 1 \times 0.947 = 1.894$$

and the factor

$$(1 - k/4) = (1 - 0.4735) = 0.5265$$

so that the eccentricity can be reduced to 0.0049 and

$$\Phi_i = 1 - 2 \times 0.0049/0.1025 = 0.90$$

Slenderness ratio
As for section (a).

Design vertical load resistance
In this section the value of $\Phi_i = 0.90$ must replace the value of $\beta = 0.78$ used in section (a) and $\gamma_m = 3.0$, resulting in a value of $30.87 f_k$ for the design vertical load resistance.

Determination of f_k
As for section (a)

$$30.87 f_k = 8.835 + 8.835 + 170.25 + (1.35 \times 17)$$
$$= 210.87\,\text{kN}$$
$$f_k = 6.83\,\text{N/mm}^2$$

Modification factors for f_k
There are no modification factors since the cross-sectional area of the wall is greater than 0.1 m^2 and the Eurocode does not include a modification factor for narrow walls.

Required value of f_k

$$f_k = 6.83\,\text{N/mm}^2 \quad \text{(compared with 8.35 in section (a))}$$

Note that in ENV 1996-1-1 an additional assumption is required for the calculation in that the modular ratio is used. This ratio is not used in BS 5628. It can be shown that for this symmetrical case the value assumed for the ratio does not have a great influence on the final value obtained for f_k. In fact for the present example taking $E_{slab}/E_{wall} = 1$ would result in $f_k = 7.0\,\text{N/mm}^2$ whilst taking $E_{slab}/E_{wall} = 4$ would result in $f_k = 6.7\,\text{N/mm}^2$.

Selection of brick/mortar combination
This selection can be achieved using the formula given in section 4.4.3.(b) Using the previously calculated value of f_k and an appropriate value for f_m, the compressive strength of the mortar, the formula can be used to find f_b, the normalized unit compressive strength. This value can then be corrected using δ, from Table 4.6, to allow for the height/width ratio of the unit used.

5.9.2 Example 2: External cavity wall (Fig. 5.17)

(a) Using BS 5628

Loads on inner leaf

	DL (kN/m)	*IL* (kN/m)
Load from above	21.1	2.2
Self-weight of wall	17.0	–
Load from slab	4.1	2.2

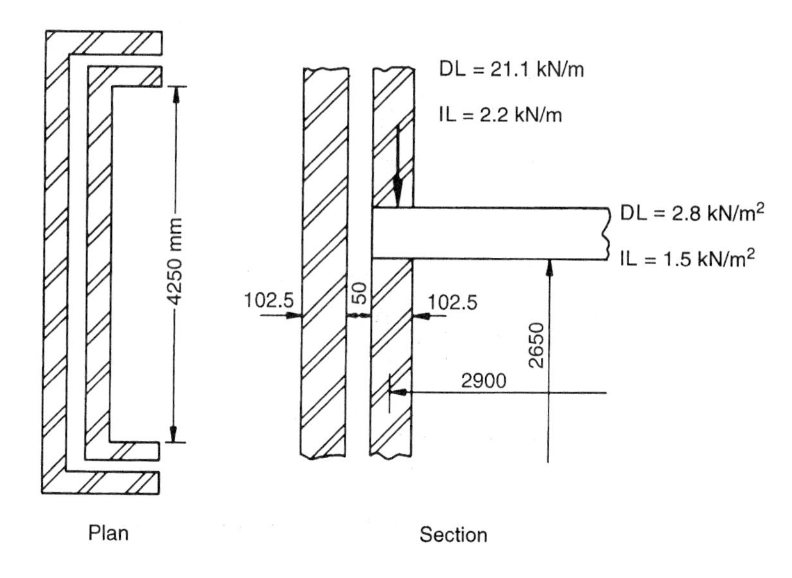

Fig. 5.17 Plan and section details for example 2.

Safety factors

$$\gamma_m = 3.5$$
$$\gamma_f \text{ (DL)} = 1.4$$
$$\gamma_f \text{ (LL)} = 1.6$$

Design vertical loading (Fig. 5.18)

Load from above $= 1.4 \times 21.1 + 1.6 \times 2.2 = 33.1$ kN/m
Self-weight of wall $= 1.4 \times 17 = 23.8$ kN/m
Total vertical design load $W_1 = 66.1$ kN/m
Load from slab $W_2 = 1.4 \times 4.1 + 1.6 \times 2.2 = 9.2$ kN/m

Slenderness ratio
Effective height $= 0.75 \times 2650 = 1988$ mm
Effective thickness $= 2(102.5 + 102.5)/3 = 136$ mm
Slenderness ratio $= 1988/136 = 14.6$

Eccentricity
See section 5.5.1. Taking moments about centre line

$$(W_1 + W_2)e_x = W_2 t/6$$
$$e_x = (9.2 \times 102.5)/6(66.1 + 9.2)$$
$$= 2.38 \text{ mm}$$

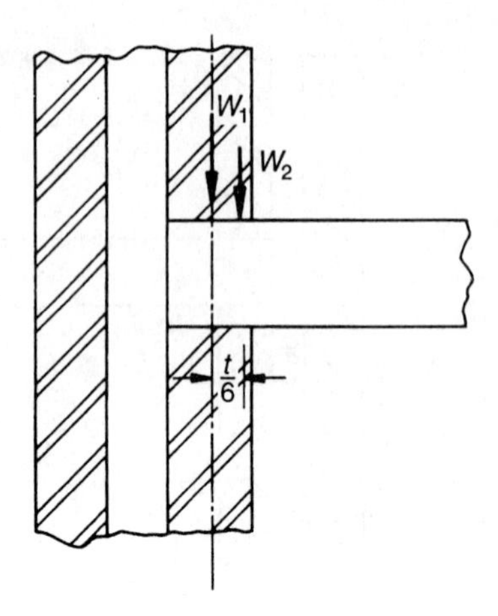

Fig. 5.18 Loading arrangement for eccentricity calculations.

From equation (5.2)

$$e_t = 0.6 \times 2.38 + 102.5\,[(1/2400) \times (14.6)^2 - 0.015]$$

$$= 1.428 + 7.566 = 8.994\,\text{mm}$$

So that, since e_t is greater than e_x, $e_m = e_t = 0.088t$ which is greater than $0.05t$, with the result that:

$$\beta = 1.1\,[1 - (2 \times 0.088)] = 0.91$$

Design vertical load resistance
Assume t in mm and f_k in N/mm^2.

$$\text{design vertical load resistance} = (\beta t f_k)/\gamma_m = 0.91 \times 102.5 \times f_k/3.5$$

$$= 26.65 f_k \quad (\text{N/mm or kN/m})$$

Determination of f_k
We have

$$\text{design vertical load} = \text{design vertical load resistance}$$

$$(33.1 + 9.2 + 23.8)\,\text{kN/m} = 26.65 f_k\,\text{kN/m}$$

$$f_k = 2.48\,\text{N/mm}^2$$

Modification factors for f_k

- Horizontal cross-sectional area $= 4.25 \times 0.1025 = 0.44\,m^2$. This is greater than $0.2\,m^2$. Therefore no modification factor for area.
- Narrow masonry wall. Wall is one brick thick; modification factor $= 1.15$.

Required value of f_k

$$f_k = 2.48/1.15 = 2.16\,N/mm^2$$

Selection of brick/mortar combination
Use Fig. 4.1 to select a suitable brick/mortar combination – nominal in this case.

(b) Using ENV 1996-1-1

The dimensions, loadings and safety factors used here are the same as those given above in section (a). The reinforced concrete floor slabs are assumed to be of the same thickness as the walls (102.5 mm) and the modular ratio E_{slab}/E_{wall} is taken as 2.

Loading
As for section (a).

Safety factors

$$\gamma_m = 3.0$$

$$\gamma_f\,(DL) = 1.35$$

$$\gamma_f\,(LL) = 1.5$$

Design vertical loading (Fig. 5.18)

Load from above $= 1.35 \times 21.1 + 1.5 \times 2.2 = 31.785\,kN/m$
Self-weight of wall $= 1.35 \times 17 = 22.95\,kN/m$
Total vertical design load $W_1 = 54.735\,kN/m$
Load from slab $W_2 = 1.35 \times 4.1 + 1.5 \times 2.2 = 8.835\,kN/m$

Eccentricity
Equation (5.8) can be rewritten:

$$M_1 = \frac{4E_w I_w/h}{8E_w I_w/h + 4E_c I_c/L_2}\left(\frac{w_2 L_2^2}{12}\right)$$

or

$$M_1 = \frac{1}{2 + (E_c I_c h / E_w I_w L_2)} \left(\frac{w_2 L_2^2}{12} \right)$$

Taking $E_c/E_w = 2$, $I_c/I_w = 1$, $h = 2650\,\text{mm}$ and the clear span $L_2 = 2797.5\,\text{mm}$

$$M_1 = \frac{1(2.8^2 \times 8.835/12)}{2 + (2 \times 1 \times 0.947)} = \frac{5.77}{3.894} = 1.48\,\text{kN m}$$

As shown in section (a)

$$N_1 = 31.785 + 8.835 + 22.95 = 63.57\,\text{kN}$$

$$M_1/N_1 = 1.48/63.57 = 0.023\,\text{m}$$

Taking $e_{hi} = 0$ and $e_a = h_{ef}/450 = 1.988/450 = 0.004\,\text{m}$ equation (5.4) becomes

$$e_i = 0.023 + 0 + 0.004 = 0.027 \quad (\geqslant 0.05t = 0.005)$$

The design vertical stress at the junction is 31.785/102.5 and since this is greater than 0.25 N/mm^2 the code allows the eccentricity to be reduced by $(1 - k/4)$ where k is given by equation 5.9.

For this example

$$k = E_c I_c h / 2E_w I_w L_2 = 2 \times 1 \times 0.947/2 = 0.947$$

and the factor

$$(1 - k/4) = (1 - 0.24) = 0.76$$

so that the eccentricity can be reduced to

$$e_i = 0.023 \times 0.76 + 0 + 0.004 = 0.0215$$

and

$$\Phi_i = 1 - 2 \times 0.0215/0.1025 = 0.58$$

Slenderness ratio
Effective height = 0.75 × 2650 = 1988 mm
Effective thickness = $(102.5^3 + 102.5^3)^{1/3}$ = 129 mm
Slenderness ratio = 1988/129 = 15.4

Design vertical load resistance
In this section the value of $\Phi_i = 0.58$ must replace the value of $\beta = 0.91$ used in section (a) resulting in a value of $19.82 f_k$ for the design vertical load resistance.

Determination of f_k
As for section (a)

$$19.82 f_k = 63.57$$
$$f_k = 3.20 \text{ N/mm}^2$$

Modification factors for f_k
There are no modification factors since the cross-sectional area of the wall is greater than 0.1 m^2 and the Eurocode does not include a modification factor for narrow walls.

Required value of f_k

$$f_k = 3.20 \text{ N/mm}^2 \quad \text{(compared with 2.16 in section (a))}$$

Note that in ENV 1996–1–1 an additional assumption is required for the calculation in that the modular ratio is used. This ratio is not used in BS 5628. It can be shown that for the present example taking $E_{slab}/E_{wall} = 1$ would result in $f_k = 4.7$ N/mm^2 whilst taking $E_{slab}/E_{wall} = 4$ would result in $f_k = 2.44$ N/mm^2. To obtain the same result from BS 5628 and ENV 1996-1-1 would require a modular ratio of 6 approximately.

Selection of brick/mortar combination
This selection can be achieved using the formula given in section 4.4.3(b) Using the previously calculated value of f_k and an appropriate value for f_m, the compressive strength of the mortar, the formula can be used to find f_b the normalized unit compressive strength. This value can then be corrected using δ, from Table 4.6, to allow for the height/width ratio of the unit used.

6

Design for wind loading

6.1 INTRODUCTION

Conventionally, in wind loading analysis, wind pressure is assumed to act statically on a structure. Such forces depend at a particular site on the mean hourly wind speed, the estimation of an appropriate gust factor, shape and pressure coefficients and the effect of local topography. The wind force calculated from these factors is assumed to act as an equivalent uniformly distributed load on the building for its full height. Sometimes the wind velocity or the gust factor is assumed variable with the height of the building, so that the intensity of the equivalent uniformly distributed load varies accordingly. In the United Kingdom, wind loads on buildings are calculated from the provisions of the Code of Practice CP 3, Chapter V, Part 2, 1970.

Whilst masonry is strong in compression, it is very weak in tension; thus engineering design for wind loading may be needed, not only for multi-storey structures, but also for some single-storey structures. Figure 6.1 shows how a typical masonry building resists lateral forces. It can be seen that two problems in wind loading design need to be considered: (1) overall stability of the building and (2) the strength of individual wall panels. In this chapter overall stability of the building will be considered.

6.2 OVERALL STABILITY

To provide stability or to stop 'card-house' type of collapse, shear walls are provided parallel to the direction of lateral loading. This is similar to diagonal bracing in a steel-framed building. In masonry structures, adequate length of walls must be provided in two directions to resist wind loads. In addition, floors must be stiff and strong enough to transfer the loads to the walls by diaphragm action. The successful action of a horizontal diaphragm requires that it should be well tied into the supporting shear walls. Section 1.2 explained in detail how lateral stability is

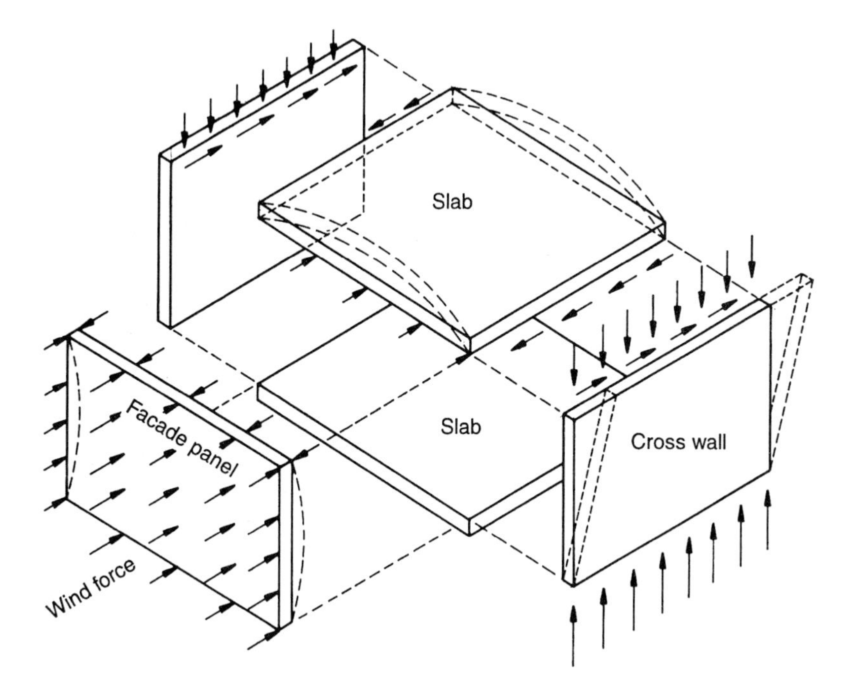

Fig. 6.1 The action of wind forces on a building. Wind force is resisted by the facade panel owing to bending, and transferred via floor slabs to the cross or shear wall and finally to the ground. (Structural Clay Products Ltd.)

provided in various types of masonry buildings, through suitable wall arrangements.

6.3 THEORETICAL METHODS FOR WIND LOAD ANALYSIS

The calculation of the lateral stiffness and stresses in a system of symmetrically placed shear walls without openings subjected to wind loading is straightforward and involves simple bending theory only. Figure 6.2 gives an illustration of such a system of shear walls.

Because of bending and shear the walls deform as cantilevers, and since the horizontal diaphragm is rigid the deflections at slab level must be the same. The deflection of individual walls is given by:

$$\Delta_1 = \frac{W_1 h^3}{3EI_1} + \frac{\lambda W_1 h}{AG} \tag{6.1}$$

$$\Delta_2 = \frac{W_2 h^3}{3EI_2} + \frac{\lambda W_2 h}{AG} \tag{6.2}$$

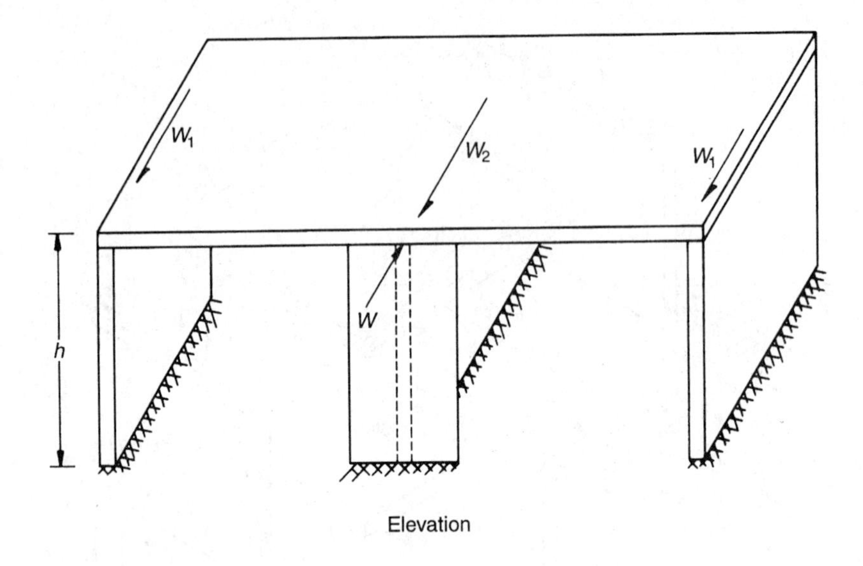

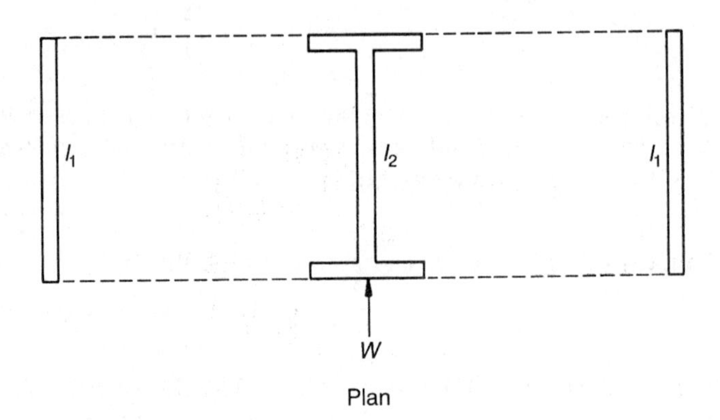

Fig. 6.2 A system of shear walls resisting wind force.

$$\Delta_1 = \Delta_2 \tag{6.3}$$

$$2W_1 + W_2 = W \tag{6.4}$$

where W_1, W_2 = lateral forces acting on individual walls, Δ_1, Δ_2 = deflections of walls, A = area of walls, h = height, E = modulus of elasticity, G = modulus of rigidity, I_1, I_2 = second moments of areas and λ = shear deformation coefficient (1.2 for rectangular section, 1.0 for flanged section).

The proportion of the lateral load carried by each wall can be obtained from equations (6.1) to (6.4). The first term in equations (6.1) and (6.2) is bending deflection and the second is that due to shear. The shear deflection is normally neglected if the height:width ratio is greater than 5.

6.3.1 Coupled shear walls

Shear walls with openings present a much more complex problem. Openings normally occur in vertical rows throughout the height of the wall, and the connection between the wall sections is provided either by beams forming the part of the wall or by floor slabs or by a combination of both. Such walls are described as 'coupled shear walls', 'pierced shear walls' or 'shear walls with openings'. Figure 6.3(a) shows a simple five-storey high coupled shear wall structure.

There are five basic methods of analysis for the estimation of wind stresses and deflection in such shear walls, namely: (i) cantilever approach, (ii) equivalent frame, (iii) wide column frame, (iv) continuum, and (v) finite element. Figures 6.3(b) to (f) show the idealization of shear walls with openings for each of these methods.

6.3.2 Cantilever approach

The structure is assumed to consist of a series of vertical cantilever walls which are made to deflect together at each level by the floor slabs. That is, the slabs transmit direct forces only, bending being neglected. The wind moment is divided amongst the walls in proportion to their flexural rigidities. This is the most commonly used method for the design of masonry structures. The deflection of the wall is given by

$$\Delta = \frac{w_1}{EI_1}\left(\frac{x^4}{24} - \frac{h^3 x}{6} + \frac{h^4}{8}\right) \tag{6.5}$$

$$\Delta = \frac{w_2}{EI_2}\left(\frac{x^4}{24} - \frac{h^3 x}{6} + \frac{h^4}{8}\right) \tag{6.6}$$

where

$$w_1 = \frac{w}{I_1 + I_2} I_1 \qquad \text{and} \qquad w_2 = \frac{w}{I_1 + I_2} I_2$$

w = total uniformly distributed wind load/unit height, h = height of building, x = distance of section under consideration from the top, and I_1, I_2 = second moments of areas (Fig. 6.3(b)).

6.3.3 Equivalent frame

In this method, the walls and slabs are replaced by columns and beams having the same flexural rigidities as the walls and floor slabs respectively. The span of the beams is taken to be the distance between the

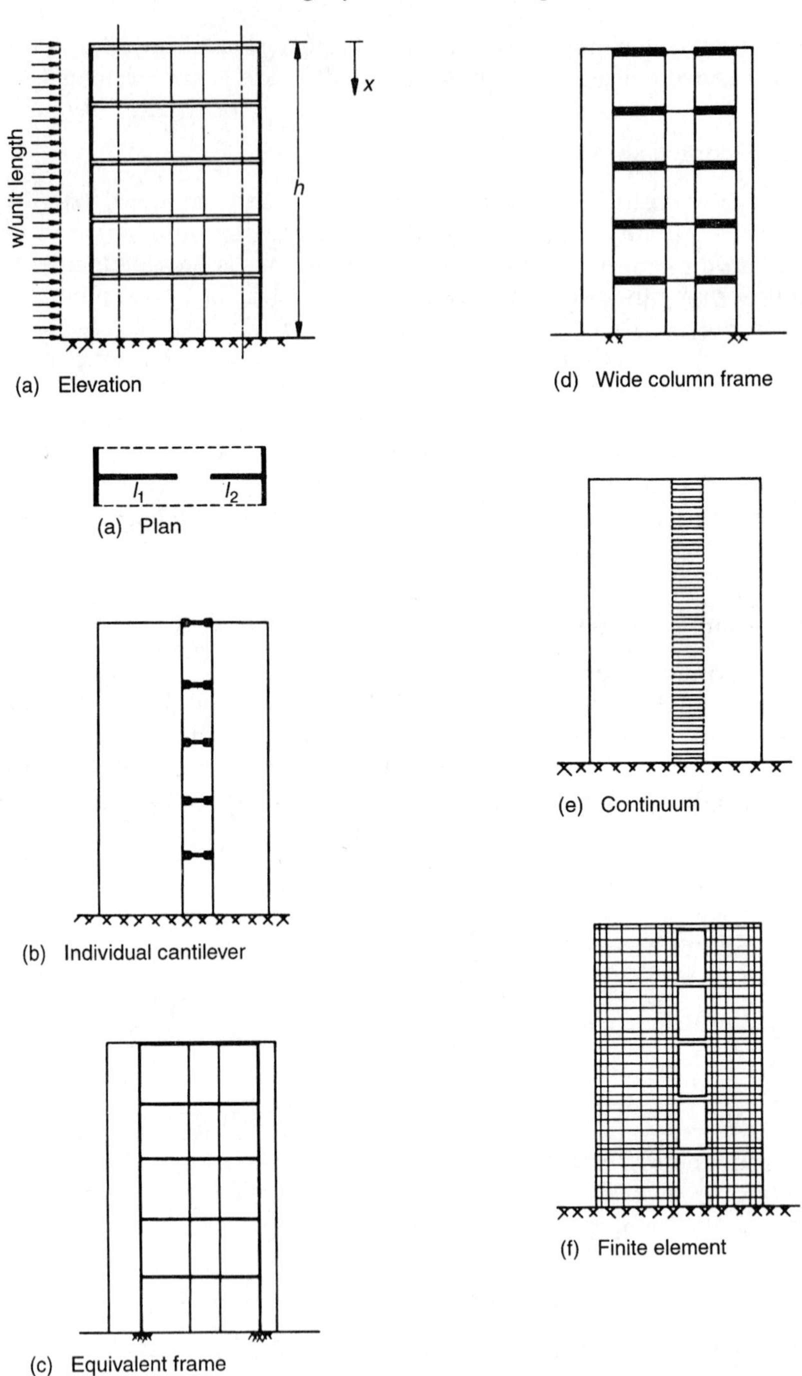

Fig. 6.3 Idealization of shear walls with opening for theoretical analysis.

centroidal axes of adjacent columns (Fig. 6.3(c)). The axial and shear deformations of beams and columns may be neglected or may be included if the structure is analysed by using any standard computer program which takes these deformations into account.

6.3.4 Wide column frame

The wide column frame is a further refinement of the equivalent frame method. The structure is idealized as in the equivalent frame method except that the interconnecting members are assumed to be of infinite rigidity for part of their length, i.e. from the centroidal axes of the columns to the opening as in Fig. 6.3(d). The system can be analysed by using a standard computer program or by conventional analysis which may or may not take into consideration the axial and shear deformation of the beams and columns.

6.3.5 Continuum

In this method, the discrete system of connecting slabs or beams is replaced by an equivalent shear medium (Fig. 6.3(e)) which is assumed continuous over the full height of the walls, and a point of contra-flexure is taken at the centre of the medium. Axial deformation of the medium and shear deformations of the walls are neglected.

Basically, the various continuum theories put forward for the analysis of a coupled shear wall are the same except for the choice of the redundant function. Readers interested are advised to consult the specialized literature for the derivation of the theory (e.g. Coull and Stafford-Smith, 1967).

6.3.6 Finite element analysis

In finite element analysis the structure is divided into a finite number of small triangular or rectangular elements (Fig. 6.3(f)), which are assumed to be connected only at their nodes. Application of the equations of equilibrium of the forces acting at these nodal points leads to a number of simultaneous equations which can be solved with the aid of a computer. The method provides a very powerful analytical tool, and suitable computer programs are readily available which can deal with any type of complex structure. However, this may prove to be a costly exercise in practical design situations.

6.3.7 Selection of analytical method

Although these methods are used in practice for analysis and design of rows of plane walls connected by slabs or beams, the analysis of a complex three-dimensional multi-storey structure presents an even more difficult problem. Furthermore, it has been observed experimentally that the results of these methods of analysis are not necessarily consistent with the behaviour of actual brick or block shear wall structures even in simple two-dimensional cases. The difference between the experimental and theoretical results may be due to the assumptions regarding interactions between the elements, which in a practical structure may not be valid because of the method of construction and the jointing materials.

To investigate the behaviour of a three-dimensional brickwork structure and the validity of the various analytical methods, a full-scale test building was built (Fig. 6.4) in a disused quarry, and lateral loads were

Fig. 6.4 (a) Test structure.

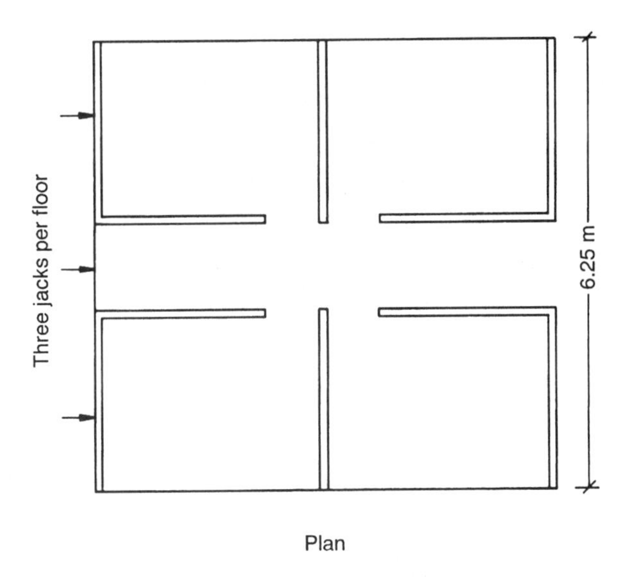

Fig. 6.4 (b) Test structure.

applied by jacking at each floor level against the quarry face, which had been previously lined with concrete to give an even working face. The deflections and strains were recorded at various loads. The three-dimensional structure was replaced by an equivalent two-dimensional wall and beam system having the same areas and moments of inertia as the actual structure and analysed by the various methods described in this chapter. The theoretical and experimental deflections are compared in Fig. 6.5. The strain and thus the stress distribution across the shear wall near ground level was nonlinear, as shown in Fig. 6.6. Most theoretical methods, with the exception of finite elements, assume a linear variation of stress across the shear wall and thus did not give accurate results. The comparisons between the various analytical methods considered (namely, simple cantilever, frame, wide column frame and shear continuum method) with experimental results strongly suggest that the best approximation to the actual behaviour of a masonry structure of this type is obtained by replacing the actual structure by an equivalent rigid frame in which the columns have the same sectional properties as the walls with interconnecting slabs spanning between the axes of the columns. The continuum or wide column frame methods do not seem to give satisfactory results for brickwork structures, and hence their use is not advisable. Finite element analysis may be justified only in special cases, and will give the nonlinear stress distribution, which cannot be reproduced by other methods.

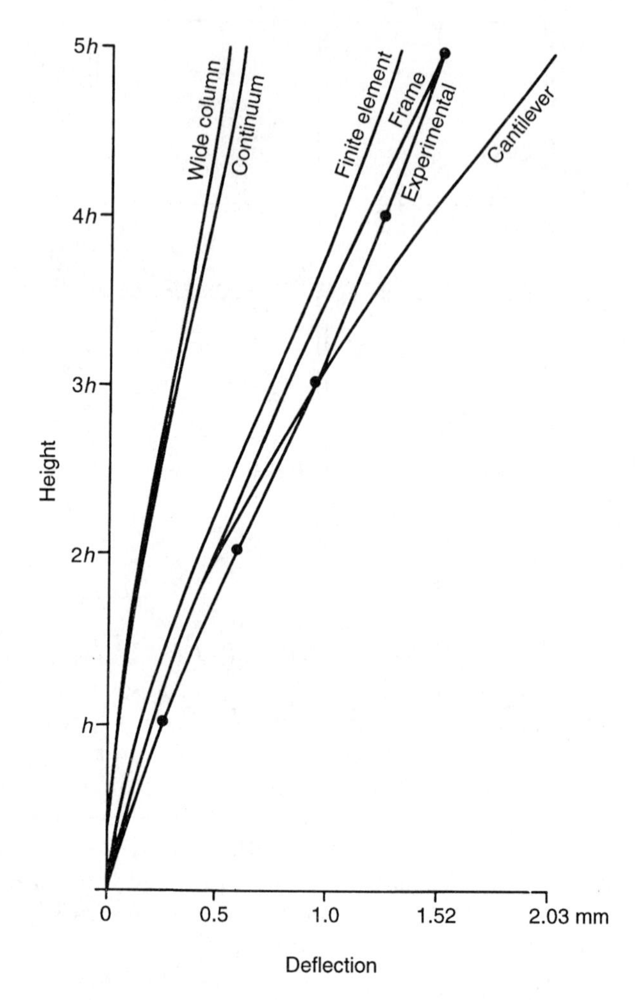

Fig. 6.5 Comparison of experimental and theoretical deflection results for an equivalent uniform load of 894 N/m^2 over the loaded face of the building.

The cantilever method of analysis is an oversimplification of the behaviour and is very conservative. For this reason, and because it is simple to carry out, it may be used for preliminary estimates of the bending moments and shearing forces in the walls of a building arising from wind loads. It should be noted, however, that this procedure neglects bending of the interconnecting beams or slabs, and this may require consideration.

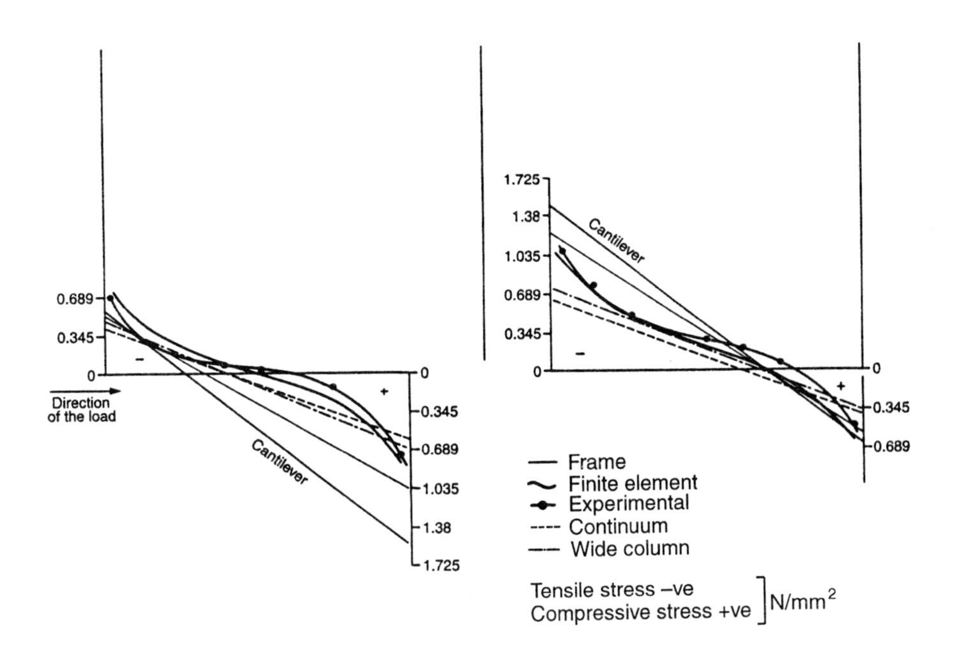

Fig. 6.6 Stress distribution across the shear walls at the base for an equivalent uniform load of 894 N/m^2 over the loaded face of the building (only one-half of the structure is shown).

6.4 LOAD DISTRIBUTION BETWEEN UNSYMMETRICALLY ARRANGED SHEAR WALLS

When a system of shear walls of uniform cross-section throughout their height is not symmetrical through either uneven spacing of walls or non-uniform distribution of mass, the resultant of the wind loads will not pass through the shear centre, i.e. the centroid of the moments of inertia, and a twisting moment will be applied to the building as illustrated in Fig. 6.7(a). Similarly, torsion will be induced in a symmetrical building, if the resultant of the applied forces does not pass through the shear centre.

The load W on the structure can be replaced by a load acting at the shear centre as in a symmetrical case, together with a twisting moment equal to We as in Fig. 6.7(b) or (c). In the case of symmetry, the load is distributed to each wall in proportion to its stiffness, since the deflection of walls must be the same at floor level. Hence

$$W_A = \frac{WI_A}{I_A + I_B + I_C} = \frac{WI_A}{\Sigma I} \quad (6.7)$$

$$W_B = \frac{WI_B}{\Sigma I}, \qquad W_C = \frac{WI_C}{\Sigma I} \quad (6.8)$$

Owing to twisting moment (We), the walls are subjected to further loading of magnitude W_A', W_B' and W_C' respectively. The loading in walls A and B will be *negative* and in wall C will be *positive*.

Assume the deflection of walls due to twisting moment is equal to Δ_a, Δ_b and Δ_c as shown in Fig. 6.8. As the floor is rigid,

$$\Delta_b = \Delta_a x_b / x_a \quad (6.9)$$

$$\Delta_c = \Delta_a x_c / x_a \quad (6.10)$$

Also

$$\Delta_a = W'_A h^3 / KEI_A \quad (6.11)$$

where K is the deflection constant and

$$\Delta_b = W'_B h^3 / KEI_B \quad (6.12)$$

Substituting the value of Δ_b from (6.9) and Δ_a from (6.11), we get

$$(\Delta_a / x_a) x_b = W_B h^3 / KEI_B$$

or

$$(x_b / x_a)(W'_A / I_A) = W_B / / I_B$$

or

$$W'_B = (W'_A I_B / I_A)(x_b / x_a) \quad (6.13)$$

Similarly,

$$W'_C = (W'_A I_C / I_A)(x_c / x_a) \quad (6.14)$$

Now the sum of the moments of all the forces about the shear centre of all the walls must be equal to the twisting moment. Hence

$$W'_A x_a + W'_B x_b + W'_C x_c = We$$

or

$$W'_A [x_a + (I_B / I_A)(x_b^2 / x_a) + (I_C / I_A)(x_c^2 / x_a)] = We$$

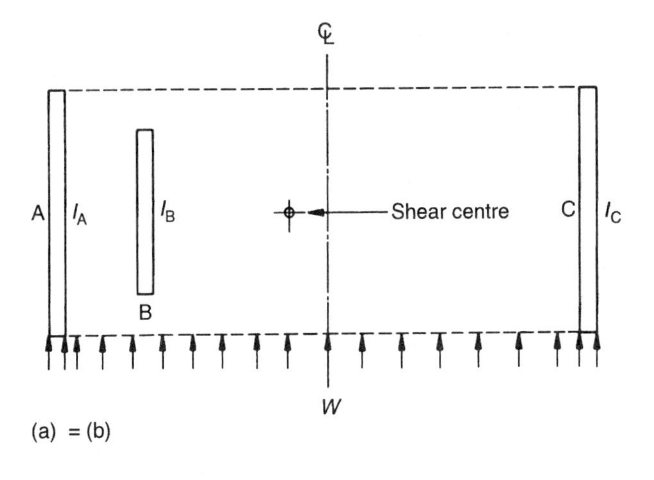

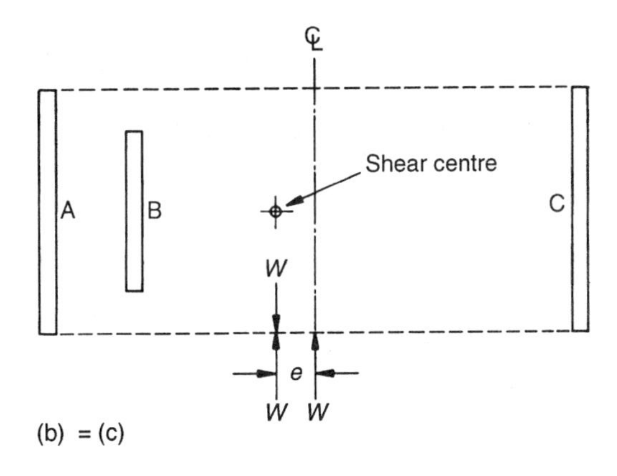

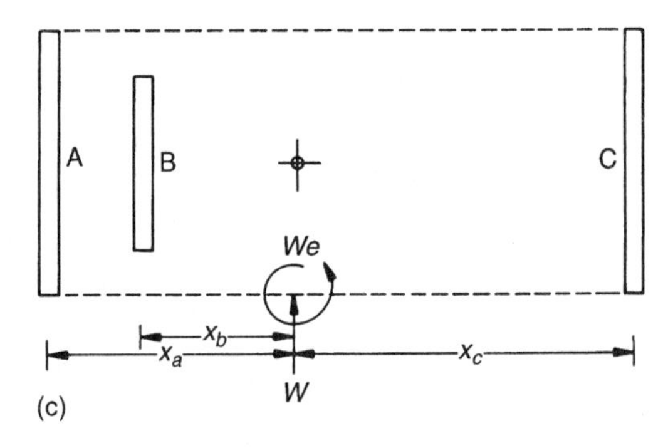

Fig. 6.7 Unsymmetrical shear walls, subjected to wind loading.

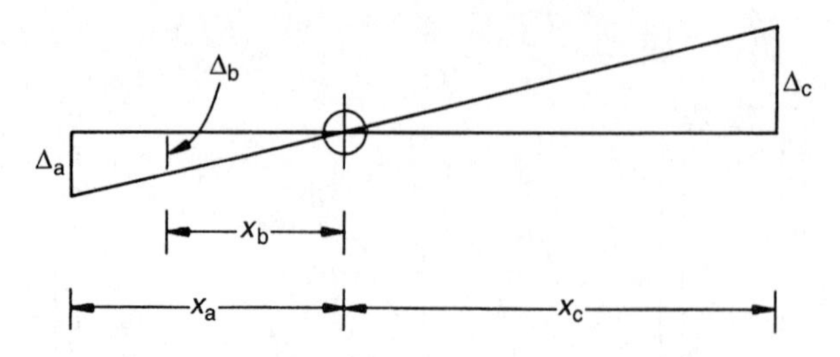

Fig. 6.8 Deflection of walls due to twisting.

or

$$W'_A = \frac{We\, I_A\, x_a}{I_A x_a^2 + I_B x_b^2 + I_C x_c^2} = \frac{We\, I_A\, x_a}{\Sigma\, I x^2} \qquad (6.15)$$

Similarly,

$$W'_B = \frac{We\, I_B\, x_b}{\Sigma\, I x^2} \quad \text{and} \quad W'_C = \frac{We\, I_C\, x_c}{\Sigma\, I x^2}$$

The load in each wall will be the algebraic sum of loads calculated from equations (6.7), (6.8) and (6.15). In other words, the load resisted by each wall can be expressed as

$$W_n = \frac{W I_n}{\Sigma I} \pm \frac{We\, I_n\, x_n}{\Sigma I x^2} \qquad (6.16)$$

The second term in the equation is positive for walls on the same side of the centroid as the load W.

7

Lateral load analysis of masonry panels

7.1 GENERAL

In any typical loadbearing masonry structure two types of wall panel resist lateral pressure, which could arise from wind forces or the effect of explosion. These panels can be classified as follows:

- Panels with precompression, i.e. panels subjected to both vertical and lateral loading.
- Panels without precompression, i.e. panels subjected to self-weight and lateral loading.

7.2 ANALYSIS OF PANELS WITH PRECOMPRESSION

The lateral strength of panels with precompression depends on the following factors:

- Flexural tensile strength
- Initial precompression
- Stiffness of the building against upward thrust
- Boundary conditions.

7.2.1 Flexural tensile strength

The flexural tensile strength of masonry normal to the bed joint is very low, and therefore it may be ignored in the lateral load design of panels with precompression without great loss of accuracy.

7.2.2 Initial precompression

As will be shown in section 7.3, the lateral strength of a wall depends on the vertical precompression applied to it. Normally this is taken to be the dead load of the structure supported by it, but if settlement occurs, it is possible for a proportion of this load to be redistributed to other parts of the structure. This is explained in simplified terms in Fig. 7.1. Relative settlement of the right-hand wall shown in the diagram will induce bending moments in the floor slabs which, in turn, will reduce the loading on this wall. The quantitative significance of this effect is shown in Fig. 7.2 which is based on measurements taken on an actual structure. As may be seen from this, relative settlement of only 1 or 2 mm can reduce the precompression by a large percentage.

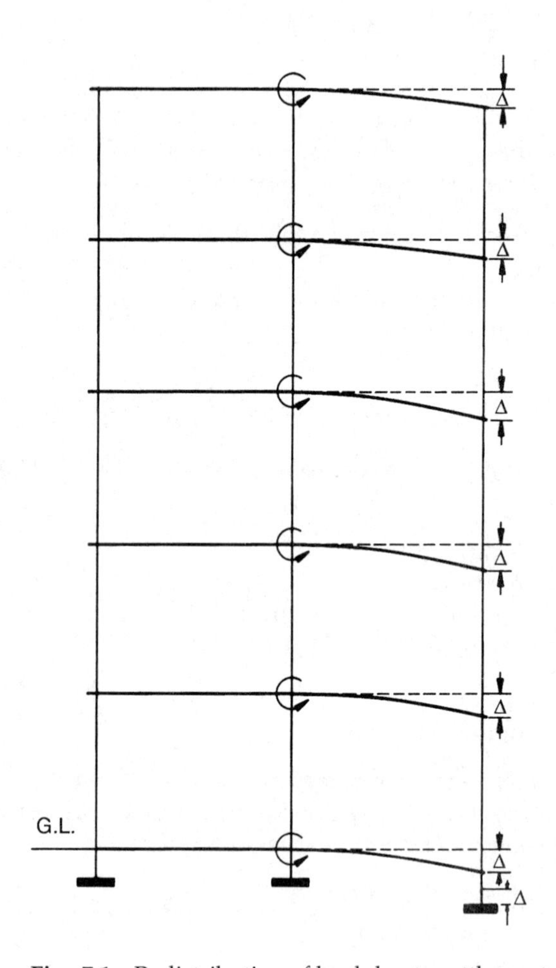

Fig. 7.1 Redistribution of load due to settlement.

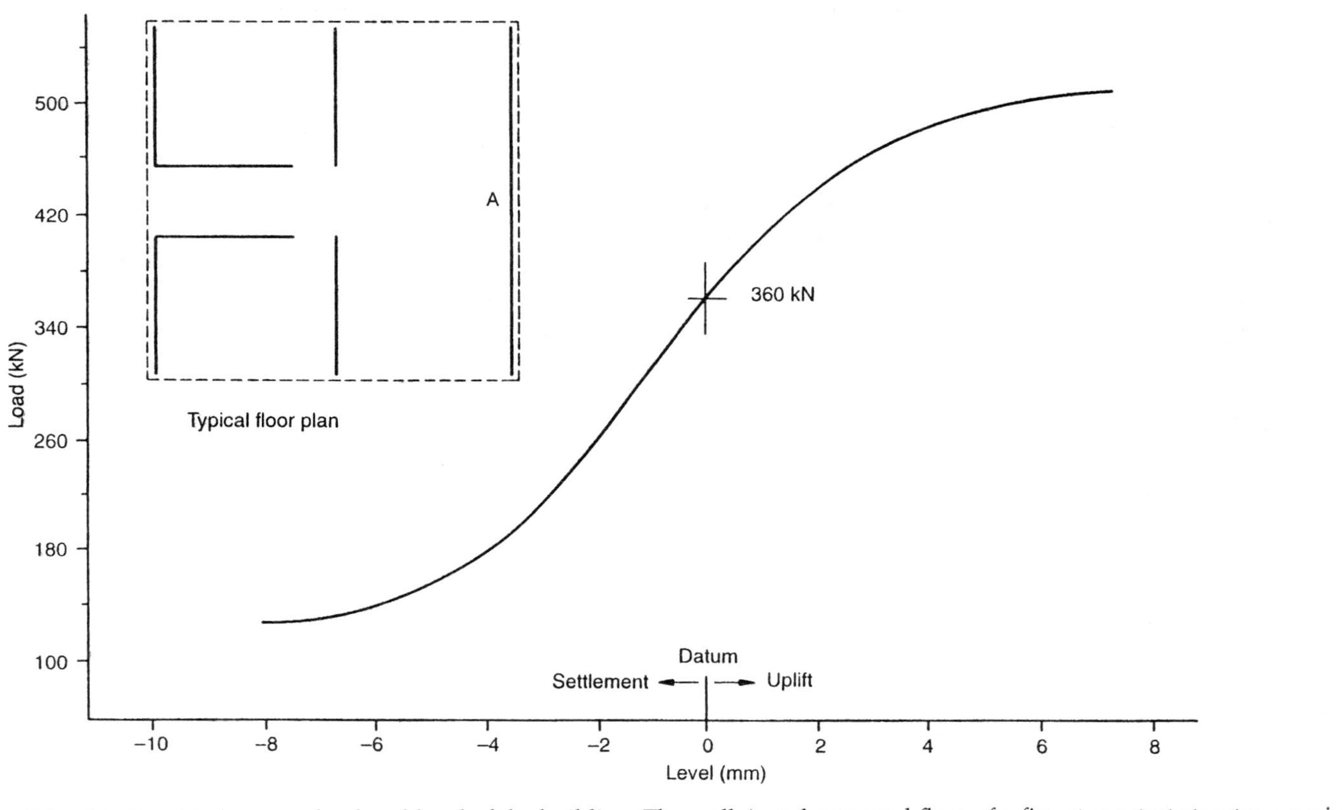

Fig. 7.2 Relationship between load and level of the building. The wall A at the ground floor of a five-storey test structure carries a load of 360 kN shown as the datum level above. When this wall is being pushed the lateral load increases owing to uplift, depending on the magnitude of uplift. In case of settlement at this end, the load carried by wall A will be reduced, depending on the magnitude of the settlement.

7.2.3 Stiffness of a building

Just before collapse a wall under lateral loading tends to lift the structure above it by a certain distance, as shown in Fig. 7.3. The uplift depends on the thickness of the wall. This is the opposite effect to that described in relation to settlement and results in an additional precompression on the wall, the value depending on the stiffness of the building against upward thrust. As shown in Fig. 7.2 the stiffness of a building, however, is highly indeterminate and nonlinear and in practical design this additional precompression may be ignored. This will add to the safety of such walls against failure under lateral pressure.

7.2.4 Boundary conditions

In practice, the walls in loadbearing masonry structures will be supported at top and bottom and may, in addition, be supported at the sides by return walls. Returns can give extra strength depending on the ratio of the length to the height of the wall attached to the return, the tensile strength of the brick or block and the number of headers tying the wall to the return. In a normal English bond alternate courses of headers are used to tie the wall to its return. In the approximate theory

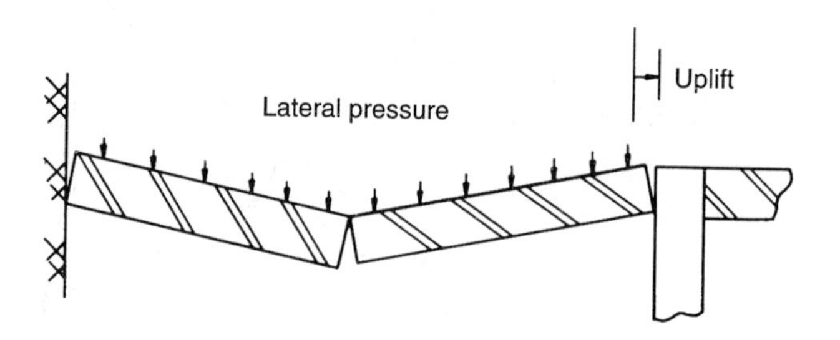

Fig. 7.3 Uplift of the slab at the time of collapse of wall beneath it.

described in the next paragraph, it has been assumed that the return does not fail. However, the designer should check whether the return can safely carry the load imposed on it.

7.3 APPROXIMATE THEORY FOR LATERAL LOAD ANALYSIS OF WALLS SUBJECTED TO PRECOMPRESSION WITH AND WITHOUT RETURNS

7.3.1 Wall without returns

Having taken into consideration all the factors contributing to the lateral strength of the wall, an approximate analysis (Hendry *et al.*, 1971) can be developed based on the following assumptions:

- Elastic deflections of the wall supports are negligible.
- Failure occurs by horizontal cracking at the top, centre and bottom of the wall, causing rotation about horizontal lines through A, B and C (Fig. 7.4).

The forces acting on the top half of the wall at the point of failure are shown in Fig. 7.4. By taking moments about A

$$\sigma t L\,(t-a) = q_0 h \frac{L}{2}\frac{h}{4} \tag{7.1}$$

$$q_0 = 8\sigma t(t-a)/h^2 \tag{7.2}$$

where σ = precompressive stress, t = thickness of the wall which is subject to precompression (in the case of a cavity wall with inner leaf loaded, thickness should be equal to the thickness of inner leaf only), L = length of wall, h = height of wall, q_0 = transverse or lateral pressure and a = horizontal distance through which centre of the wall has moved.

If the compressive stress is assumed constant throughout the uplift of the wall at failure, the maximum pressure resisted by the wall is equal to

$$q_0 = 8\sigma t^2/h^2 \quad \text{when} \quad a = 0 \tag{7.3}$$

If the precompression increases on the wall with uplift of the building, as explained above, it is possible for the moment of resistance, $\sigma t L\,(t-a)$, to increase, even though the moment arm $(t-a)$ decreases – thus resulting in an increase in the maximum lateral pressure resisted by the wall.

7.3.2 Wall with returns

In the case of a wall with returns, part of the lateral pressure is transmitted to the return, thus causing axial and bending stresses in the return.

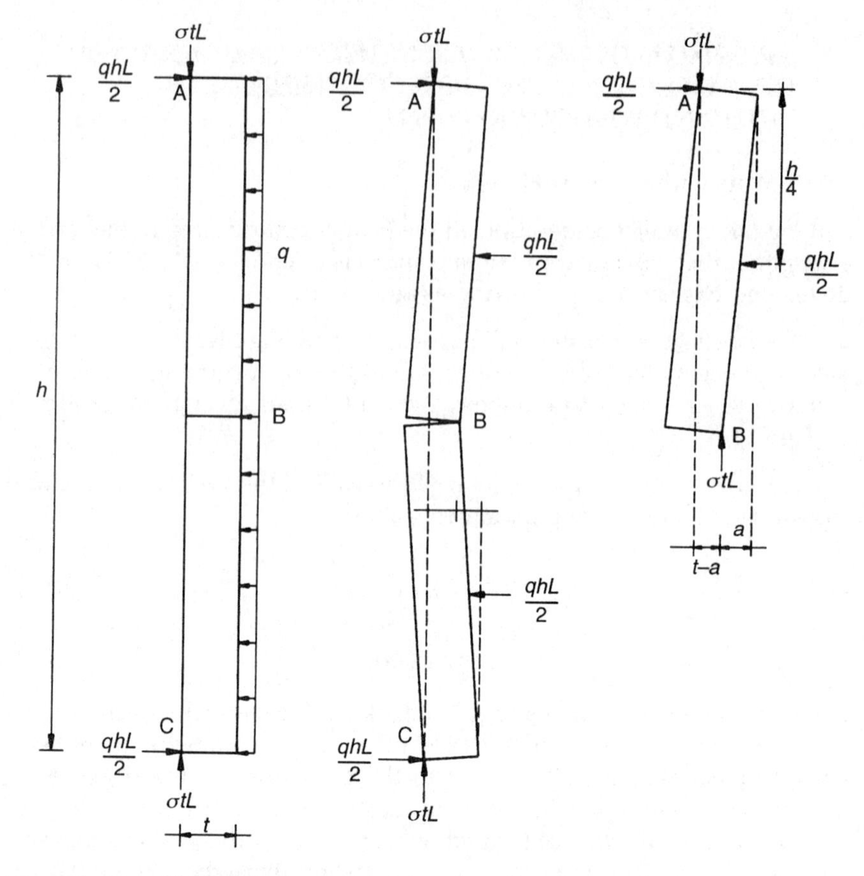

Fig. 7.4 Simplified failure mechanism of walls supported top and bottom. q = lateral pressure; σ = precompressive stress; L = length of wall.

In simplified analysis, however, the return is assumed not to fail. The lateral pressure transmitted to the return is assumed to be distributed over the height of the wall at 45° (Fig. 7.5).

Considering a wall with one return and taking the moment of all the forces acting on the top half of the wall (Fig. 7.5) about the top:

$$\sigma t^2 L = q_1 \frac{h}{2} L \frac{h}{4} - q_1 \frac{h^2}{8} \frac{h}{3} \tag{7.4}$$

Therefore

$$q_1 = \frac{8\sigma t^2}{h^2} \frac{1}{1 - 1/(3\alpha)} \quad \text{where} \quad \alpha = L/h \geqslant 0.5 \tag{7.5}$$

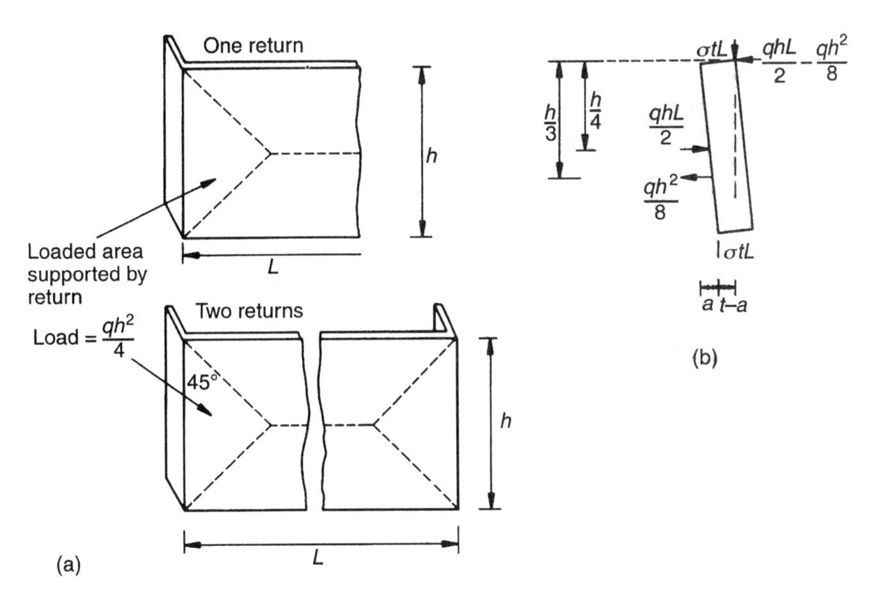

Fig. 7.5 Simplified failure mechanism for walls with returns.

Now substituting the value of q_0 (wall with no return) from equation (7.3) into equation (7.5)

$$q_1 = q_0 \frac{1}{1 - 1/(3\alpha)} \tag{7.6}$$

Similarly, for a wall with two returns (Fig. 7.5 (a)):

$$\sigma t^2 L = q_2 \frac{hL}{2}\frac{h}{4} - q_2 \frac{h^2}{8}\frac{2h}{3} \tag{7.7}$$

$$q_2 = \frac{8\sigma t^2}{h^2}\frac{1}{1 - 2/(3\alpha)} \quad \text{where} \quad \alpha = L/h \geqslant 1.0 \tag{7.8}$$

From equation (7.3)

$$q_2 = q_0 \frac{1}{1 - 2/(3\alpha)} \tag{7.9}$$

For various values of α, the q_1/q_0 and q_2/q_0 plots have been shown in Fig. 7.6 together with the experimental results.

In the British Code of Practice BS 5628 the factors $1/[1 - 1/(3\alpha)]$ and $1/[1 - 2/(3\alpha)]$ are replaced by a single factor k. Table 7.1 shows the comparison between factor k obtained from the theory and from the code. From Table 7.1 it can be seen that the British code values are in good agreement with the theoretical results. The theoretical values in

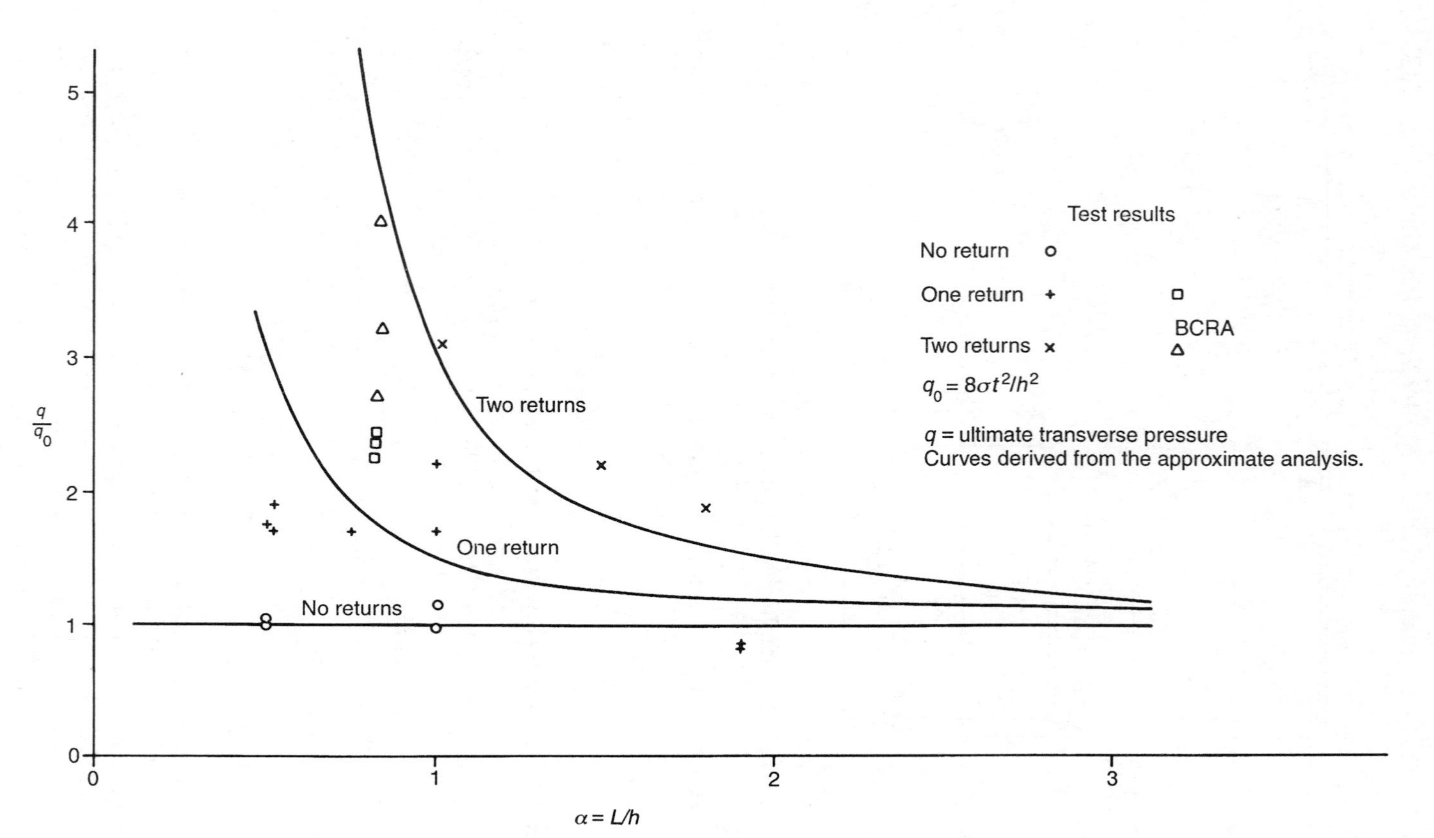

Fig. 7.6 Effect of returns on the lateral strength of walls with varying L/h ratios.

Table 7.1 Comparison of the value k

Number of returns	*Value of k*			
	$L/h = 0.75$	1.0	2.0	3.0
1	1.6(1.8)[a]	1.5(1.5)	1.1(1.2)	1.0(1.1)
2	4.0 (–)	3.0(3.0)	1.5(1.5)	1.2(1.28)

[a]Theoretical values in brackets.

some cases have been slightly adjusted in the light of experimental results which are shown in Fig. 7.6. Because of the simplified assumption that the return will not fail before the wall, the curves (Fig. 7.6) for q_1/q_0 and q_2/q_0 at lower L/h ratio become asymptotic to the Y-axis which is physically not possible; hence the code has used a cut-off point on the evidence of experimental results.

7.4 EFFECT OF VERY HIGH PRECOMPRESSION

From equation (7.3) it can be seen that the lateral pressure varies directly with precompression; this is perfectly true for an ideal rigid body. In masonry walls with high precompression, as the two blocks rotate (Fig. 7.7) on top of each other resulting in a reduced effective cross-sectional area with very high local stress approaching ultimate strength in crushing, the failure will be earlier than predicted by the straight-line theory of equation (7.3). At a precompression equal to the ultimate strength of masonry, the wall will fail without resisting any lateral pressure. From Fig. 7.8, which has been derived analytically taking into account the deformation of the wall, it can be seen that the maximum capacity of resisting lateral pressure for any strength of brickwork is reached at a precompression equal to approximately half of the ultimate strength. After this value of precompression, the lateral load-resisting capacity of a wall decreases. As the design stress in compression utilizes only a fraction of the ultimate masonry strength and will never exceed half of the ultimate strength, in almost all practical cases the failure condition will be in the linear range of Fig. 7.8; hence the simplified approximate analysis can accurately and safely be applied. This may also be assumed in the case of blockwork walls subjected to precompression.

7.5 LATERAL LOAD DESIGN OF PANELS WITHOUT PRECOMPRESSION

Masonry panels which resist out-of-plane lateral loading may be supported as follows:

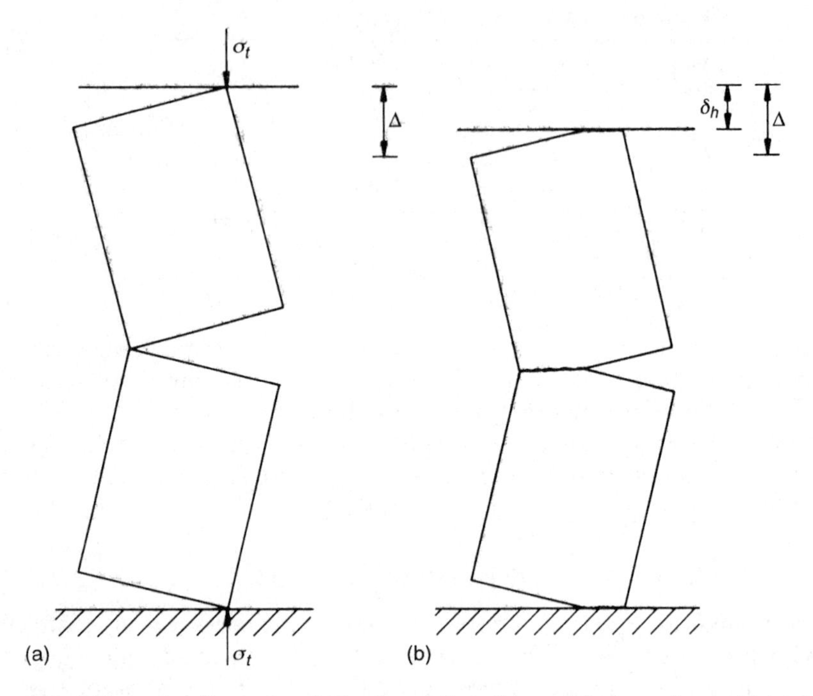

Fig. 7.7 Effect of wall rotation: (a) basic rotation; (b) modified rotation (with high precompression). σ = precompression; Δ = half maximum uplift of wall with no corner deformations; δ_h = elastic shortening.

- Simply supported top and bottom, i.e. vertically spanning panel.
- Simply supported on two edges, i.e. horizontally spanning panel.
- Simply supported or continuous on three or four sides, i.e. panels supported on more than two sides of various boundary conditions.

It will of course be realized that simple supports are an idealization of actual conditions which will usually be capable of developing some degree of moment resistance.

7.5.1 Vertically or horizontally spanning panels

The maximum moments per unit width for a wall spanning vertically or horizontally can be calculated from:

vertically spanning panel

$$M_y = wh^2/8 \tag{7.10}$$

horizontally spanning panel

$$M_x = wL^2/8 \tag{7.11}$$

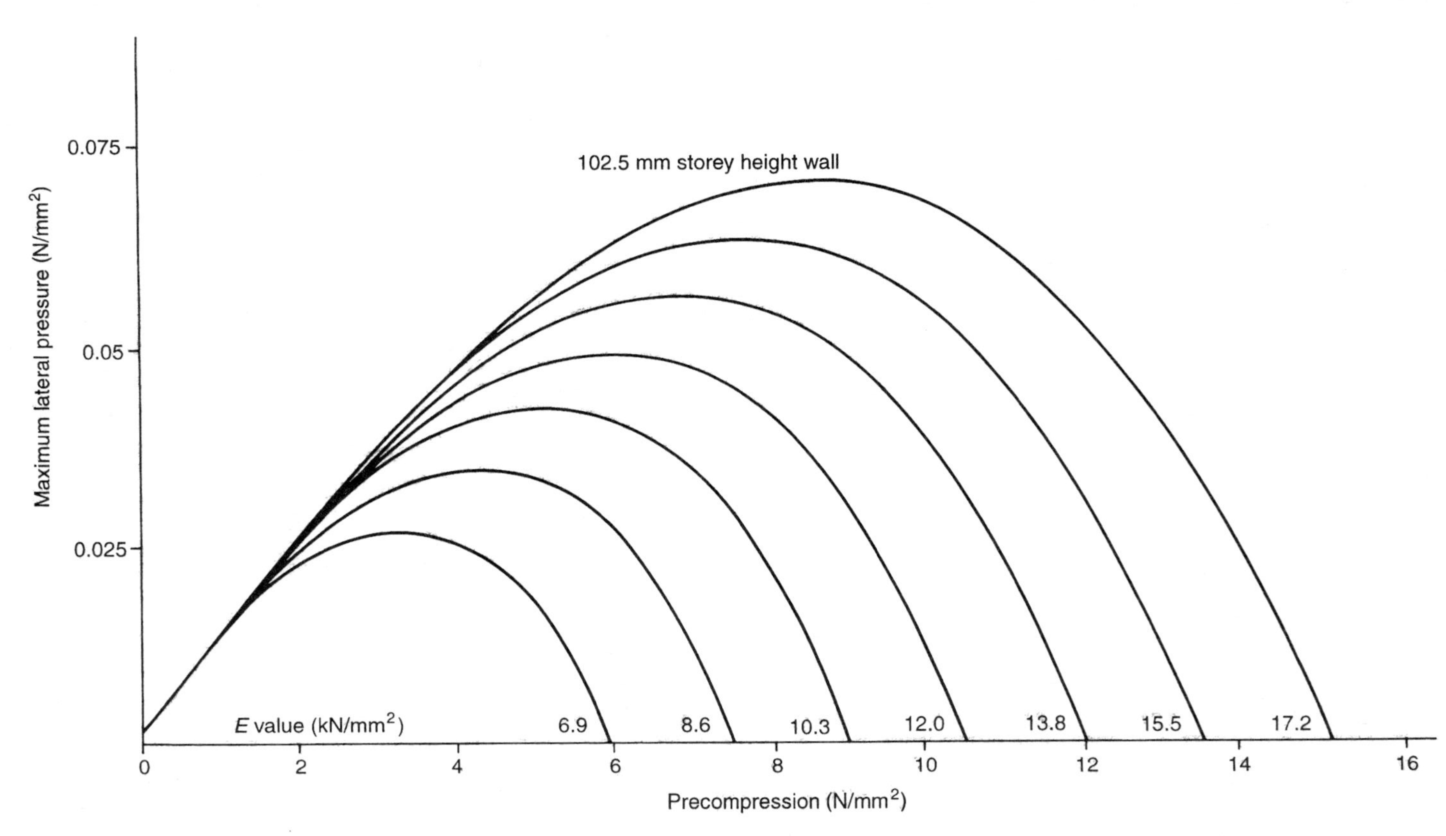

Fig. 7.8 Precompression versus maximum lateral pressure on 102.5 mm wall of storey height.

where w = design pressure, M_x and M_y = maximum moments per unit width at midspan on strips of unit width and span h and L.

Similarly, the moment of resistance per unit width of the panel can be calculated from the known value of the flexural tensile strengths in respective directions as:

$$M_y = f_{ty} Z \tag{7.12}$$

$$M_x = f_{tx} Z \tag{7.13}$$

where f_{ty} = allowable tensile strength perpendicular to the bed joint, f_{tx} = allowable tensile strength parallel to the bed joint and Z = sectional modulus for unit width.

In case of limit state design, the design bending moments per unit width in two directions will be

$$M_y = w_k \gamma_f h^2/8 \tag{7.14}$$

$$M_x = w_k \gamma_f L^2/8 \tag{7.15}$$

where w_k = characteristic wind load per unit area and γ_f = partial safety factor for loads.

The moment of resistance of the panel spanning vertically and horizontally will be given by

$$M_y = f_{ky} Z/\gamma_m \tag{7.16}$$

$$M_x = f_{kx} Z/\gamma_m \tag{7.17}$$

where f_{ky} and f_{kx} are characteristic tensile strength normal and parallel to bed joints.

7.5.2 Panels supported on more than two sides with various boundary conditions

The lateral load analysis of masonry panels of various boundary conditions is very complicated since masonry has different strength and stiffness properties in two orthogonal directions. Some typical values of brickwork moduli of elasticity on which the stiffness depends are given in Table 7.2.

The British limit state code BS 5628 recommends bending moment coefficients for the design of laterally loaded panels. The code does not indicate the origin of these coefficients, but they are numerically equal to those given by yield-line analysis as applied to under-reinforced concrete slabs with corresponding boundary conditions. Strictly speaking, yield-line analysis is not applicable to a brittle material like masonry which cannot develop constant-moment hinges as occur in reinforced

Table 7.2 Moduli of elasticity of brickwork in two orthogonal directions (Grade I and II mortar)

Type of brick	*Modulus of elasticity of brickwork*	
	E_y (kN/mm^2)	E_x (kN/mm^2)
Single frog (low strength, 21.55 N/mm^2)	6.2	8.8
Double frog (medium strength, 59.40 N/mm^2)	8.65	15.3
Three holes perforated (high strength, 88.33 N/mm^2)	18.13	16.53

concrete with yielding of the steel. It is not surprising, therefore, that a comparison between test results and those derived from yield-line analysis shows that the yield-line method consistently overestimates the failure pressure of brickwork panels when the orthogonal ratio is interpreted as the strength ratio. Since brickwork or blockwork panels exhibit different strengths and stiffness properties in two orthogonal directions, a simplified method for the design based on fracture lines taking into account both strength and stiffness orthotropies is discussed below. The method has been applied to predict the failure pressure of rectangular panels, rectangular panels with opening, and octagonal and triangular panels of various boundary conditions, and may be used for the design of brickwork or blockwork panels using the published values of the stiffness orthotropy and flexural strengths.

7.5.3 Fracture-line analysis

The fracture-line analysis which is described here is an ultimate load design method for laterally loaded panels. For more details see Sinha (1978, 1980).

Assumptions

All deformations take place along the fracture lines only, and the individual parts of the slab rotate as rigid bodies. The load distribution is in accordance with the stiffnesses in the respective directions. The fracture lines develop only when the relevant strengths are reached simultaneously in both directions.

Consider the idealized fracture lines for a four-sided panel with two simply supported and two continuous edges (see Fig. 7.9). Every portion of the panel into which it is divided by the fracture lines is in equilibrium under the action of external forces and reactions along the fracture lines and supports.

Since it is symmetrical, only parts 1 and 2 need consideration. In case of asymmetry the entire rigid area needs to be considered.

Consider triangle AFB:

$$\text{load on AFB}\,(1) = \tfrac{1}{2} w\beta\alpha L^2 \tag{7.18}$$

and its moment along AB is

$$\text{moment} = \tfrac{1}{2} w\beta\alpha L^2 x(\beta\alpha L/3) \quad \text{(since CG of load drops 1/3)}$$

$$= w\beta\alpha^2 L^3/6 \tag{7.19}$$

For equilibrium

$$w\beta^2\alpha^2 L^3/6 = mL$$

Therefore

$$w\beta\alpha^2 L^2/6 = m/\beta \tag{7.20}$$

Similarly, for AFED (2) (the left-hand side of equation (7.21) has been obtained by dividing the rigid body 2 into two triangles and one rectangle for simplification of the calculation)

$$(w\beta L^2/12) + (wL^2/8) - (w\beta L^2/4) = 2\mu m/K \qquad \text{where} \quad K = E_x/E_y$$

$$(wL^2/12)\;(\beta + 1.5 - 3\beta) = 2\mu m/K$$

$$(w\alpha^2 L^2/6)\;(1.5 - 2\beta) = 4\mu m\alpha^2/K \tag{7.21}$$

From equations (7.20) and (7.21),

$$(w\alpha^2 L^2/6)\;(1.5 - 2\beta + \beta) = (m/\beta) + (4\mu m\alpha^2/K) \tag{7.22}$$

or

$$(w\alpha^2 L^2/6)\;(1.5 - \beta) = (m/\beta)\;[1 + (4\mu m\alpha^2/K)]$$

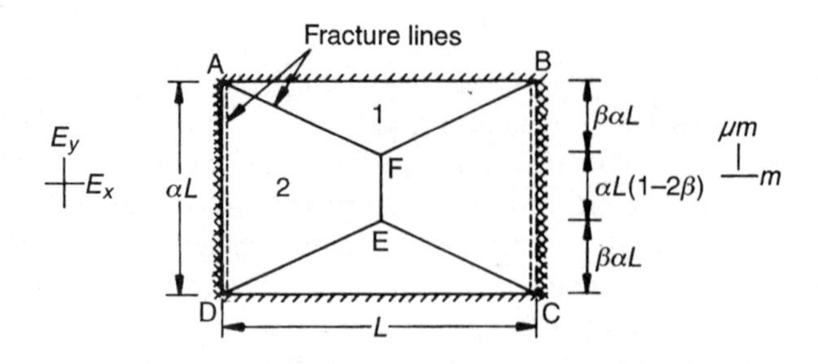

Fig. 7.9 Idealized fracture lines.

therefore

$$m = \frac{w\alpha^2 L^2}{6}\left(\frac{1.5\beta - \beta^2}{1 + (4\mu\beta\alpha^2/K)}\right) \tag{7.23}$$

For minimum collapse load or maximum value of moment $d(m/w)/d\beta = 0$, from which

$$\beta = \frac{K}{4\mu\alpha^2}\left[\left(\frac{6\mu\alpha^2}{K} + 1\right)^{1/2} - 1\right] \tag{7.24}$$

The value of β can be substituted in equations to obtain the relationship between the failure moment and the load.

For a particular panel, the fracture pattern that gives the lowest collapse load should be taken as failure load. The values of m and β for various fracture-line patterns for panels of different boundary conditions are given in Table 7.3, and the reader can derive them from first principles as explained above.

7.5.4 How to obtain the bending-moment coefficient of BS 5628 or EC6 from the fracture-line analysis

Although the fracture-line method has been suggested for accurate analysis, the designer may prefer to use the BS 5628 coefficients. Hence this section briefly outlines the method to obtain the coefficients from the fracture line. In BS 5628 the bending-moment coefficients are given for horizontal bending (M_x), whereas the analysis presented in this chapter considers the vertical bending (M_y).

Similarly, the orthotropy ratio in case of BS 5628 is taken as the ratio

$$\frac{\text{strength normal to bed joint}}{\text{strength parallel to bed joint}}$$

Hence the orthotropy is less than 1, whereas in the present analysis the orthotropy is the reciprocal of this ratio.

The BS 5628 coefficients can be obtained by putting $K = 1$ in equations (7.24) and (7.23) and also in the equation of Table 7.3, and by multiplying the vertical moment (M_y) by the orthotropy defined as in the fracture-line analysis. The provisions of EC 6 for lateral load design for resistance to wind loads are the same as BS 5628, and hence need no separate explanation.

Example

Consider the case of a panel similar to Fig. 7.9. We have

$$\frac{\text{strength parallel to bed joint}}{\text{strength normal to bed joint}} = 3.33; \quad \alpha = h/L = 0.75$$

Table 7.3 Ultimate moment for panels of different boundary conditions.

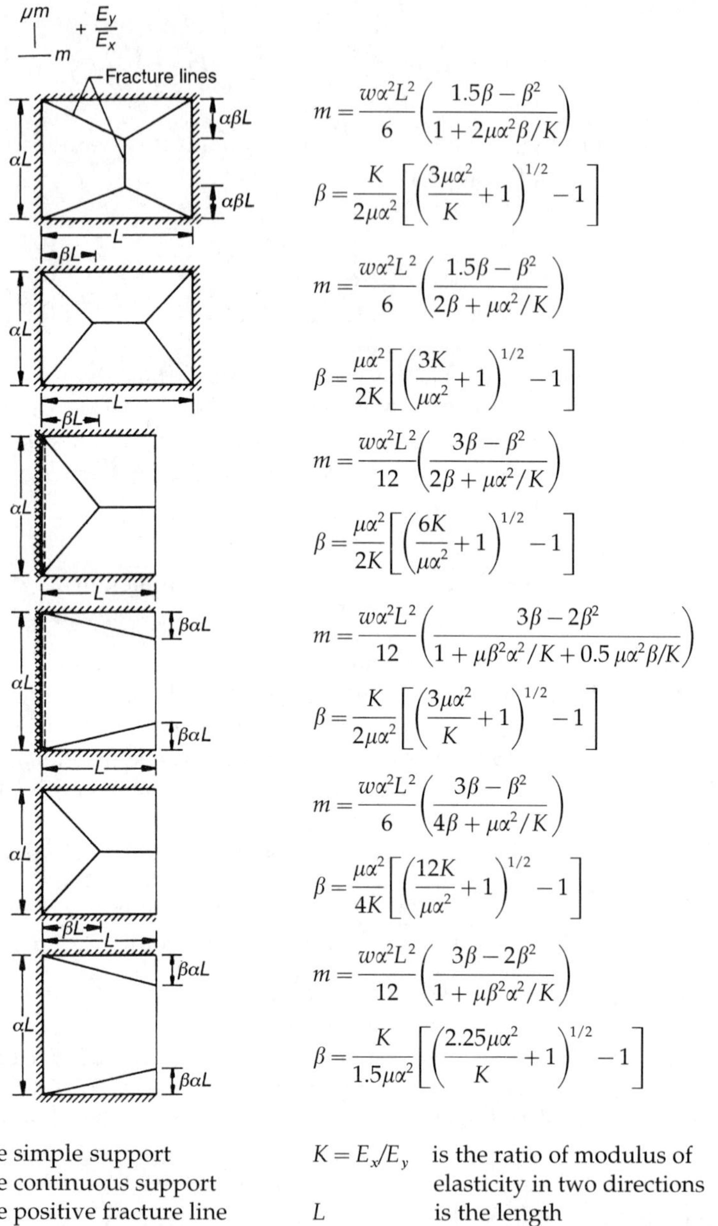

$$m=\frac{w\alpha^2L^2}{6}\left(\frac{1.5\beta-\beta^2}{1+2\mu\alpha^2\beta/K}\right)$$

$$\beta=\frac{K}{2\mu\alpha^2}\left[\left(\frac{3\mu\alpha^2}{K}+1\right)^{1/2}-1\right]$$

$$m=\frac{w\alpha^2L^2}{6}\left(\frac{1.5\beta-\beta^2}{2\beta+\mu\alpha^2/K}\right)$$

$$\beta=\frac{\mu\alpha^2}{2K}\left[\left(\frac{3K}{\mu\alpha^2}+1\right)^{1/2}-1\right]$$

$$m=\frac{w\alpha^2L^2}{12}\left(\frac{3\beta-\beta^2}{2\beta+\mu\alpha^2/K}\right)$$

$$\beta=\frac{\mu\alpha^2}{2K}\left[\left(\frac{6K}{\mu\alpha^2}+1\right)^{1/2}-1\right]$$

$$m=\frac{w\alpha^2L^2}{12}\left(\frac{3\beta-2\beta^2}{1+\mu\beta^2\alpha^2/K+0.5\,\mu\alpha^2\beta/K}\right)$$

$$\beta=\frac{K}{2\mu\alpha^2}\left[\left(\frac{3\mu\alpha^2}{K}+1\right)^{1/2}-1\right]$$

$$m=\frac{w\alpha^2L^2}{6}\left(\frac{3\beta-\beta^2}{4\beta+\mu\alpha^2/K}\right)$$

$$\beta=\frac{\mu\alpha^2}{4K}\left[\left(\frac{12K}{\mu\alpha^2}+1\right)^{1/2}-1\right]$$

$$m=\frac{w\alpha^2L^2}{12}\left(\frac{3\beta-2\beta^2}{1+\mu\beta^2\alpha^2/K}\right)$$

$$\beta=\frac{K}{1.5\mu\alpha^2}\left[\left(\frac{2.25\mu\alpha^2}{K}+1\right)^{1/2}-1\right]$$

Notation

(hatched line)	is the simple support
(cross-hatched line)	is the continuous support
———	is the positive fracture line
--------	is the negative fracture line
m	is the ultimate moment/ unit length along the bed joint
μm	is the ultimate moment/unit length normal to bed joint
$K = E_x/E_y$	is the ratio of modulus of elasticity in two directions
L	is the length
α	is the height/length ratio (h/L)
w	is the failure pressure
β	is a factor

(Note that in BS 5628 the symbol α is used for bending moment coefficient.)

From equation (7.24)

$$\beta = \frac{K}{4\mu\alpha^2}\left[\left(\frac{6\mu\alpha^2}{K}+1\right)^{1/2}-1\right]$$

$$= \frac{1}{4 \times 3.33 \times (0.75)^2}\left[\left(\frac{6 \times 3.33 \times (0.75)^2}{1}+1\right)^{1/2}-1\right] = 0.3334$$

From equation (7.23) vertical moment

$$m = \frac{w\alpha^2 L^2}{6}\left(\frac{1.5\beta-\beta^2}{1+4\mu\beta\alpha^2/K}\right)$$

$$= \frac{wL^2}{6} \times (0.75)^2\left(\frac{1.5 \times 0.3334-(0.3334)^2}{1+4 \times 3.33 \times 0.3334 \times (0.75)^2/1}\right) = 0.0104\, wL^2$$

therefore horizontal moment

$$\mu m = 0.0104 \times 3.33 wL^2 = 0.035 wL^2$$

The bending moment coefficient from BS 5628 for the corresponding case ($hlL = 0.75$) is also 0.035.

8

Composite action between walls and other elements

8.1 COMPOSITE WALL–BEAMS

8.1.1 Introduction

If a wall and the beam on which it is supported can be considered to act as a single composite unit then, for design purposes, the proportion of the load acting on the wall which is carried by the supporting beam must be determined. Prior to 1952 it was common practice to design the beams or lintels so as to be capable of carrying a triangular load of masonry in which the span of the beam represented the base of an equilateral triangle. The method allowed for a proportion of the self-weight of the masonry but ignored any additional superimposed load.

Since that period a great deal of research, both practical and theoretical, has been undertaken, and a better understanding of the problem is now possible.

Consider the simply supported wall–beam shown in Fig. 8.1. The action of the load introduces tensile forces in the beam due to the bending of the deep composite wall–beam and, since the beam now acts as a tie, the supports are partially restrained horizontally so that an arching action results in the panels. The degree of arching is dependent on the relative stiffness of the wall to the beam, and it will be shown later that both the flexural stiffness and the axial stiffness must be taken into account. In general, the stiffer the beam the greater the beam-bending moment since a larger proportion of the load will be transmitted to the beam.

The values of the vertical and horizontal stresses depend on a number of factors, but typical plots of the vertical and horizontal stress distributions along XX and YY of Fig. 8.1 are shown in Fig. 8.2.

Note that the maximum vertical stress, along the wall–beam interface, occurs at the supports and that at mid-span the horizontal

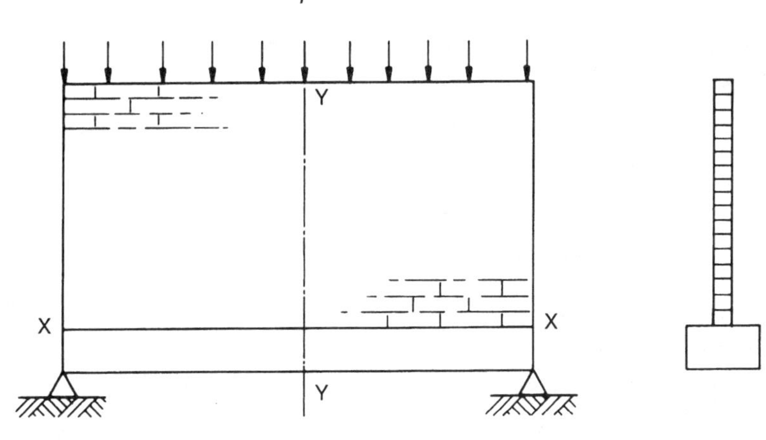

Fig. 8.1 Simply supported wall–beam.

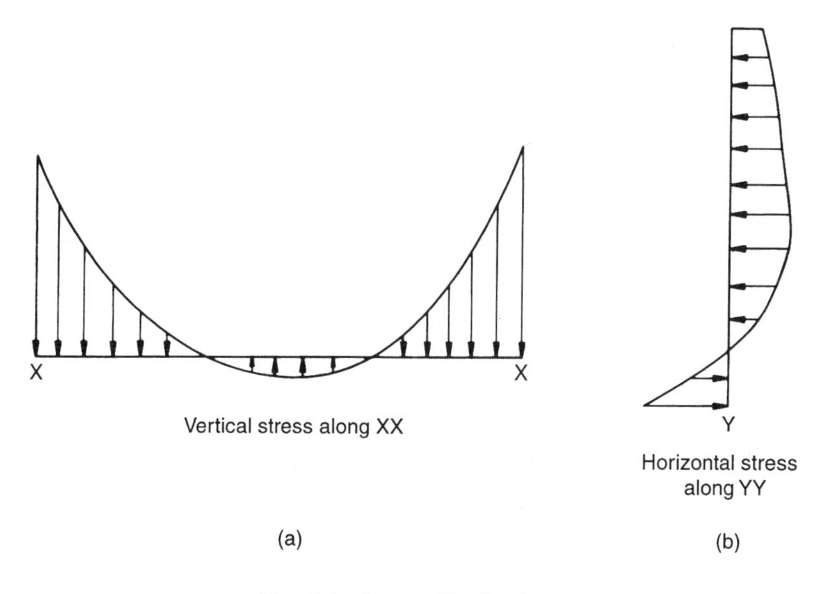

Fig. 8.2 Stress distribution.

stresses in the beam may be tensile throughout the depth so that the beam acts as a tie.

Composite action cannot be achieved unless there is sufficient bond between the wall and the beam to allow for the development of the required shearing forces. The large compressive stresses near the supports result in large frictional forces along the interface, and it has been shown that if the depth/span ratio of the wall is > 0.6 then the frictional forces developed are sufficient to supply the required shear capacity.

8.1.2 Development of design methods

For design purposes the quantities which must be determined are:

- The maximum vertical stress in the wall.
- The axial force in the beam.
- The maximum shear stress along the interface.
- The central bending moment in the beam.
- The maximum bending moment in the beam and its location.

Methods which allowed for arching action were developed by Wood (1952) for determining the bending moment and axial force in the beams. The panels were assumed to have a depth/span ratio greater than 0.6 so that the necessary relieving arch action could be developed and moment coefficients were introduced to enable the beam bending moments to be determined. These were:

- $PL/100$ for plain walls or walls with door or window openings occurring at centre span.
- $PL/50$ for walls with door or window openings occurring near the supports.

An alternative approach, based on the assumption that the moment arm between the centres of compression and tension was 2/3 × overall depth with a limiting value of 0.7 × the wall span (Fig. 8.3) was also suggested (Wood and Simms, 1969). Using this assumption, the tensile force in the beam can be calculated using

$$T \times 2h/3 = PL/8 \tag{8.1}$$

and the beam designed to carry this force.

Following this early work of Wood and Simms, the composite wall–beam problem was studied by a number of researchers who considered not only the design of the beam but also the stresses in the wall. The characteristic parameter K introduced by Stafford-Smith and Riddington (1977) to express the relative stiffness of the wall and beam was shown to be a useful parameter for the determination of both the compressive stresses in the wall and the bending moments in the beam. The value of K is given by

$$K = (E_w tL^3/E_{bm}I_b)^{1/4} \tag{8.2}$$

where E_w, E_{bm} = Young's moduli of the wall and beam respectively, I_b = second moment of area of the beam and t, L = wall thickness and span. The parameter K does not contain the variable h since it was

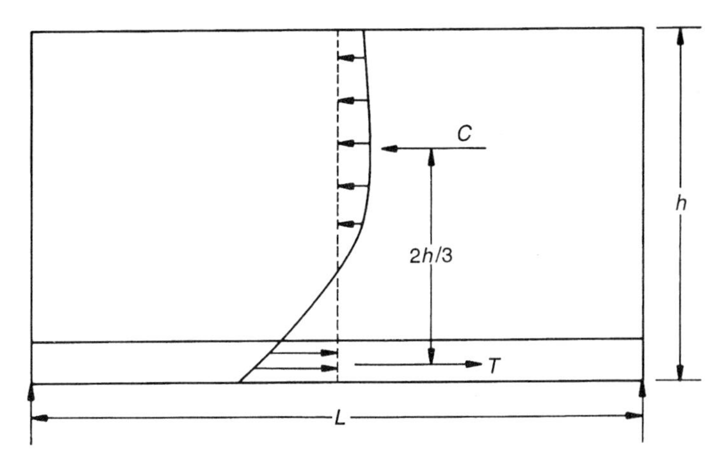

Fig. 8.3 Moment capacity of wall–beam.

considered that the ratio of h/L was equal to 0.6 and that this was representative of walls for which the actual h/L value was greater.

Conservative estimates of the stresses in walls on beam structures with restrained or free ends based on the above are:

$$\text{maximum moment in beam} = PL/4\,(E_w tL^3/E_{bm}I_b)^{1/3} \tag{8.3}$$

$$\text{maximum tie force in beam} = P/3.4 \tag{8.4}$$

$$\text{maximum stress in wall} = 1.63\,(P/Lt)\,(E_w tL^3/E_{bm}I_b)^{0.28} \tag{8.5}$$

Note that assuming $h/L = 0.6$, equation (8.1) above becomes $T = P/3.2$ which is similar to equation (8.4).

In 1980 an approximate method of analysis based on a graphical approach was introduced. This method is described in section 8.1.4.

8.1.3 Basic assumptions

The walls considered are built of brickwork or blockwork and the beams of concrete or steel. It is assumed that there is sufficient bond between the wall and the beam to carry the shear stress at the interface, and this presupposes that a steel beam would be encased and the ratio of h/L would be $\geqslant 0.6$.

The loading, including the self-weight of the wall, is represented by a istributed load along the top surface. Care must be taken with additional loads placed at beam level since the tensile forces that might result could destroy the composite action by reducing the frictional resistance.

Two stiffness parameters, R and K_1, are introduced to enable the appropriate stresses and moments to be determined. The first is a flexural stiffness parameter similar to that introduced by Stafford-Smith and Riddington (1977) except that the height of the wall replaces the span, and the second is an axial stiffness parameter used for determining the axial force in the beam:

$$R = (E_w th^3 / E_{bm} I_b)^{1/4} \tag{8.6}$$

$$K_1 = E_w th / E_{bm} A_b \tag{8.7}$$

A typical vertical stress distribution at the wall–beam interface is shown in Fig. 8.2(a). To simplify the analysis it is assumed that the distribution of this stress can be represented by a straight line, a parabola or a cubic parabola depending on the range of R shown in Fig. 8.4.

The axial force in the beam is assumed to be linear with a maximum value at the centre and zero at the supports.

8.1.4 The graphical method

(a) Maximum vertical stress in wall (f_m)

This stress is a maximum over the supports and can be determined using the equation

$$f_m = (P/L\,t)\,C_1 \tag{8.8}$$

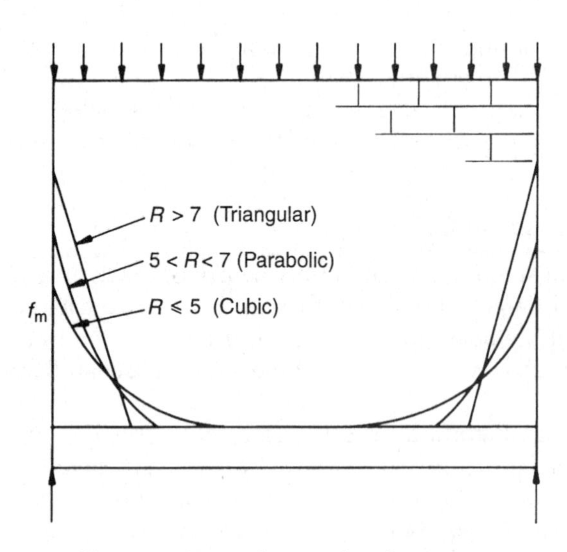

Fig. 8.4 Vertical stress distribution.

where C_1 can be obtained from Fig. 8.5 using the calculated values of R and h/L.

(b) Axial force in the beam (T)

This force is assumed to be a maximum at the centre and can be determined using the equation

$$T = PC_2 \tag{8.9}$$

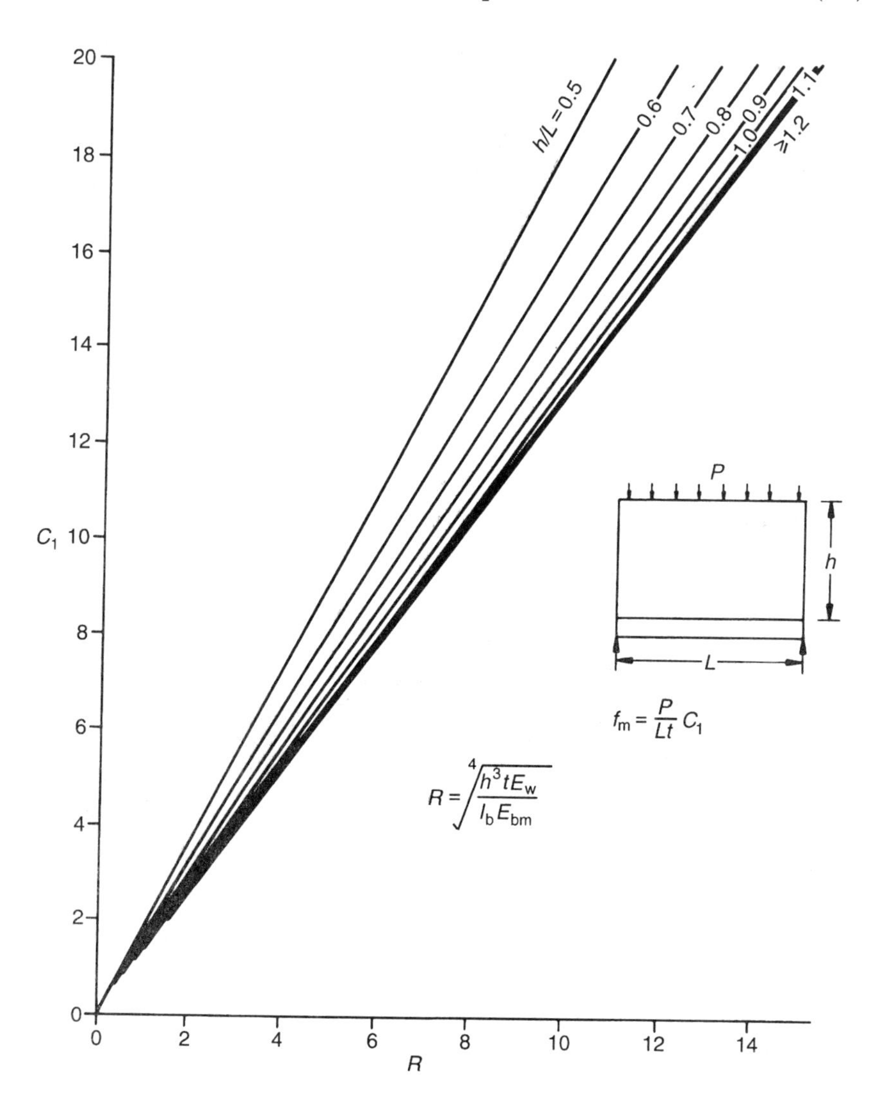

Fig. 8.5 Flexural stiffness parameter.

where C_2 can be found from Fig. 8.6 using the calculated values of K_1 and h/L.

(c) Maximum shear stress along interface (τ_m)

The maximum interface shear occurs near the supports and can be determined using

$$\tau_m = (P/Lt)\,C_1C_2 \tag{8.10}$$

where C_1 and C_2 are the values already obtained from Figs 8.5 and 8.6.

(d) Bending moments in the beam

The maximum bending moment in the beam does not occur at the centre, because of the influence of the shear stresses along the interface. Both the maximum and central bending moments can, however, be

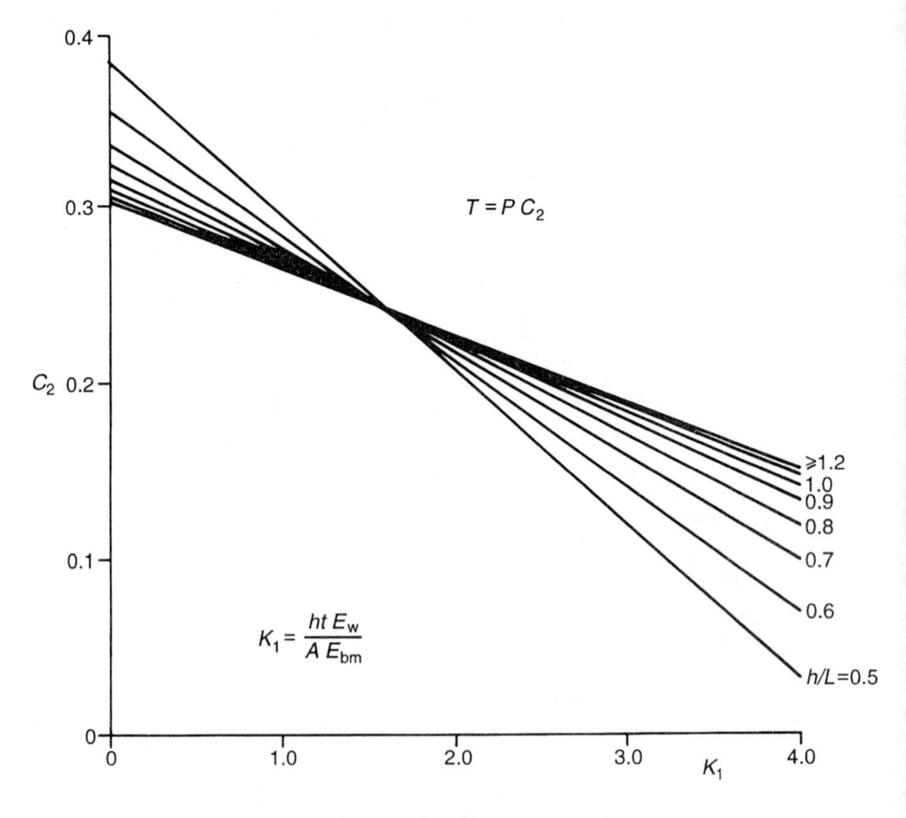

Fig. 8.6 Axial stiffness parameter.

obtained from one graph (for a particular range of R) by using the appropriate abscissae. The three graphs, Figs. 8.7, 8.8 and 8.9, have been drawn so that each represents a relationship for the particular range of R shown.

To obtain the maximum moment, the lower C_1 scale is used and for the central moment the $C_1 \times C_2$ scale is used. In each case use of the appropriate d/L ratio will give the value of

$$MC_1/PL \tag{8.11}$$

where M is either the maximum or the central bending moment.

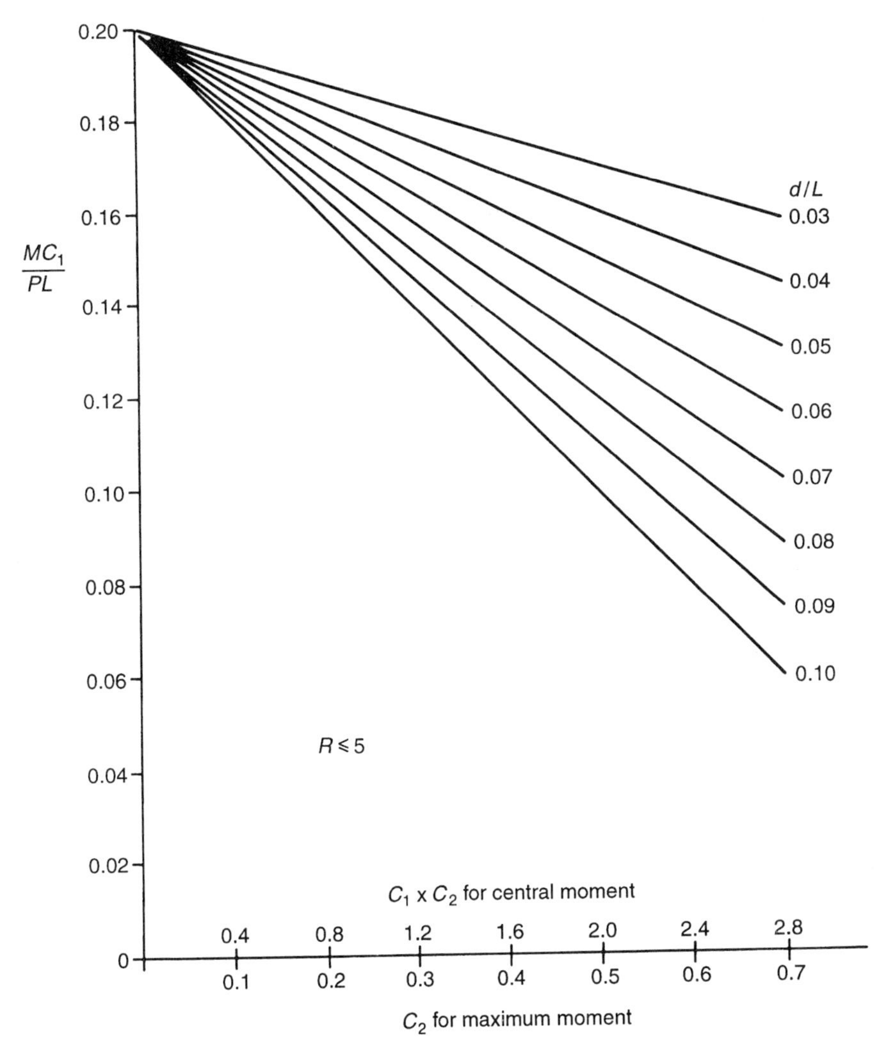

Fig. 8.7 Moments for cubic stress distribution.

The location of the maximum moment is not so important for design purposes but if required an approximate value can be determined from the equation

$$l = P/(2Sf_{m}t) \tag{8.12}$$

where S is a coefficient which depends on the shape of the vertical stress diagram and can be assumed to be

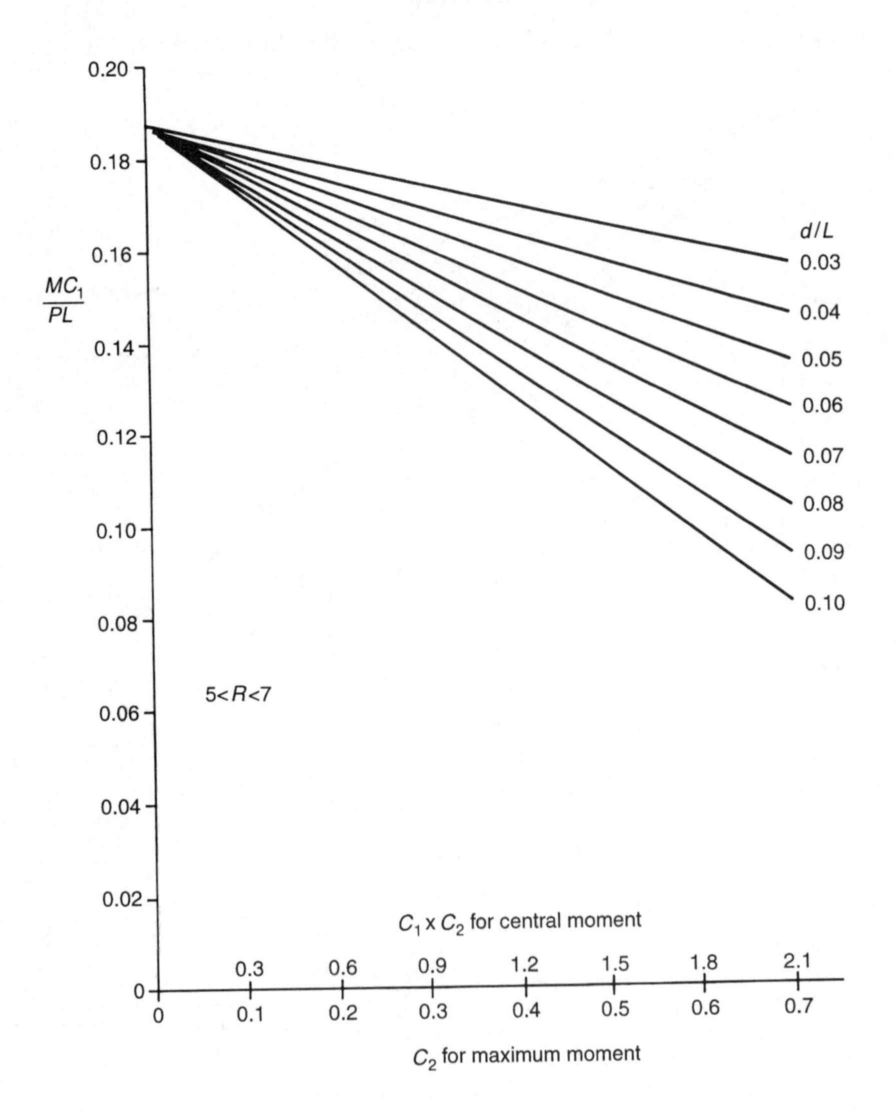

Fig. 8.8 Moments for parabolic stress distribution.

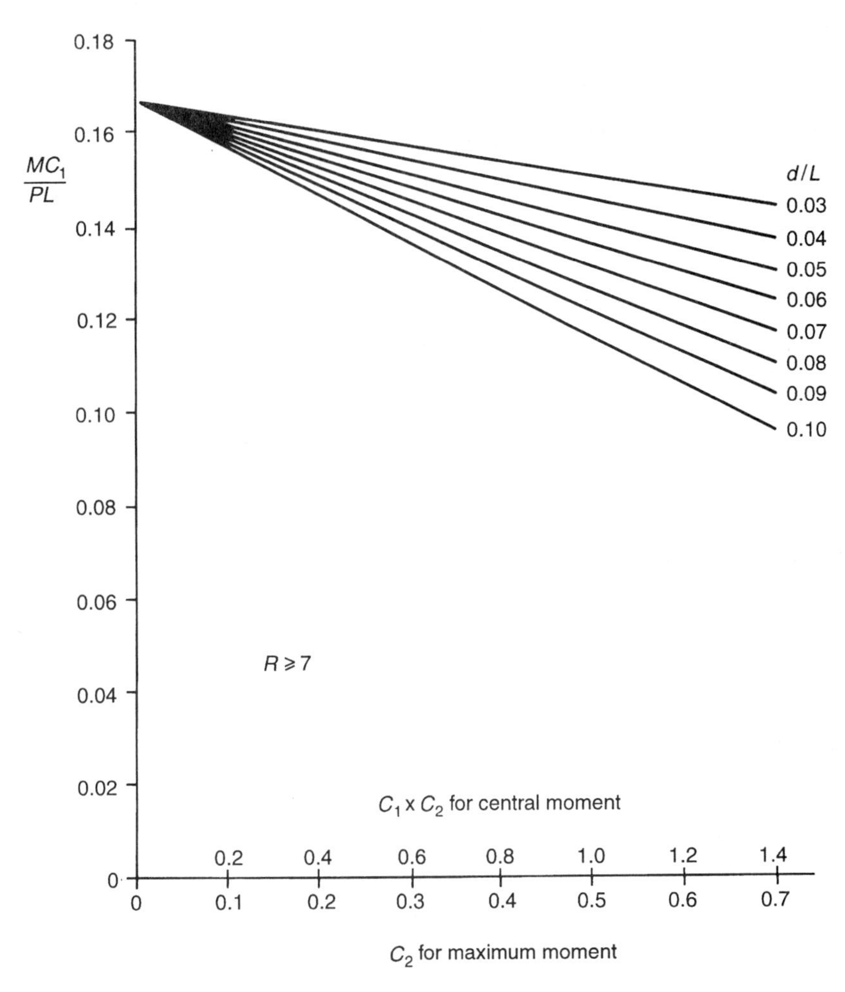

Fig. 8.9 Moments for triangular stress distribution.

- $S = 0.30$ for $R \leqslant 5$
- $S = 0.33$ for R between 5 and 7
- $S = 0.5$ for $R \geqslant 7$

(e) Example

To illustrate the use of the method consider the wall–beam shown in Fig. 8.10. Here

$$E_{bm}/E_w = 30$$

$$I_b = 115 \times 218/12 = 9.93 \times 10^7\,\text{mm}^4$$

$$R = [(1829^3 \times 115 \times 1)/(9.93 \times 10^7 \times 30)]^{1/4} = (236.23)^{1/4} = 3.92$$

$$K_1 = (1829 \times 115)/(30 \times 115 \times 218) = 0.28$$

$$h/L = 1829/2743 = 0.67$$

$$d/L = 218/2743 = 0.079$$

$$\text{total load} = 0.07 \times 2743 \times 115 = 22\,081\,\text{N}$$

Using the graphs, $C_1 = 6.8$ and $C_2 = 0.325$. Therefore

$$f_m = (22\,081 \times 6.8)/(2743 \times 115) = 0.476\,\text{N/mm}^2 \quad (8.13)$$

$$T = 22\,081 \times 0.325 = 7176.3\,\text{N} \quad (8.14)$$

$$\tau_m = (22\,081 \times 6.8 \times 0.325)/(2743 \times 115) = 0.1547\,\text{N/mm}^2 \quad (8.15)$$

From Fig. 8.7, $M_c C_1/PL = 0.115$ and $M_m C_1/PL = 0.144$ where M_c = centre line moment and M_m = maximum moment, or

$$M_c = (0.115 \times 22\,081 \times 2743)/6.8 = 1.02 \times 10^6\,\text{N mm}$$

$$M_m = (0.144 \times 22\,081 \times 2743)/6.8 = 1.28 \times 10^6\,\text{N mm}$$

Location of maximum moment from support

$$22\,081/(2 \times 0.3 \times 0.48 \times 115) = 666.7\,\text{mm}$$

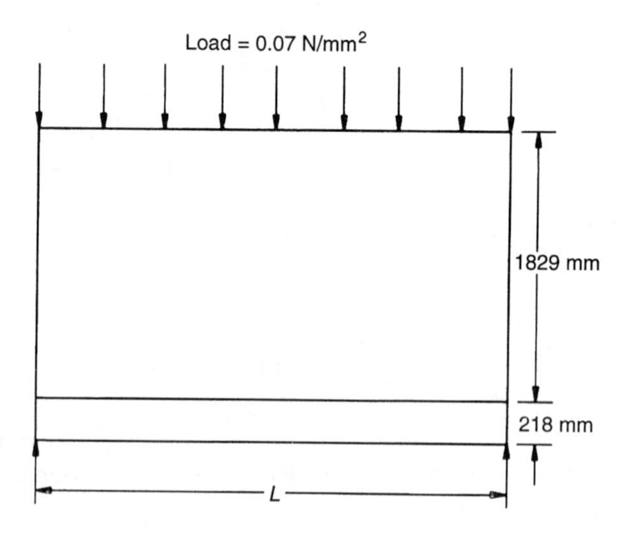

Fig. 8.10 Dimensions for wall beam example. $L = 2743$ mm, $b = t = 115$ mm.

These calculations are carried out in terms of design loads and are to be compared with the design strengths of the material in compression and shear. The design of the beam would be carried out in accordance with the relevant code of practice.

8.2 INTERACTION BETWEEN WALL PANELS AND FRAMES

8.2.1 Introduction

Wall panels built into frameworks of steel or reinforced concrete contribute to the overall stiffness of the structure, and a method is required for predicting modes of failure and calculating stresses and lateral collapse loads.

The problem has been studied by a number of authors, and although methods of solution have been proposed, work is still continuing and more laboratory or field testing is required to verify the proposed theoretical approaches.

A theoretical analysis based on a fairly sophisticated finite element approach which allowed for cracking within the elements as the load was increased was used by Riddington and Stafford-Smith (1977). An alternative method developed by Wood (1978) was based on idealized plastic failure modes and then applying a correcting factor to allow for the fact that masonry is not ideally plastic.

These methods are too cumbersome for practical design purposes, and simplifying assumptions are made for determining acceptable approximate values of the unknowns.

The basis of the design method proposed by Riddington and Stafford-Smith is that the framed panel, in shear, acts as a diagonal strut, and failure of the panel occurs owing to compression in the diagonal or shear along the bedding planes. The beams and columns of the frame are designed on the basis of a simple static analysis of an equivalent frame with pin-jointed connections in which panels are represented as diagonal pin-jointed bracing struts.

A description of the design method proposed by Wood is given below.

8.2.2 Design method based on plastic failure modes

(a) Introduction

In the method proposed by Wood (1978) four idealized plastic failure modes are considered, and these together with the location of plastic hinges are shown in Fig. 8.11.

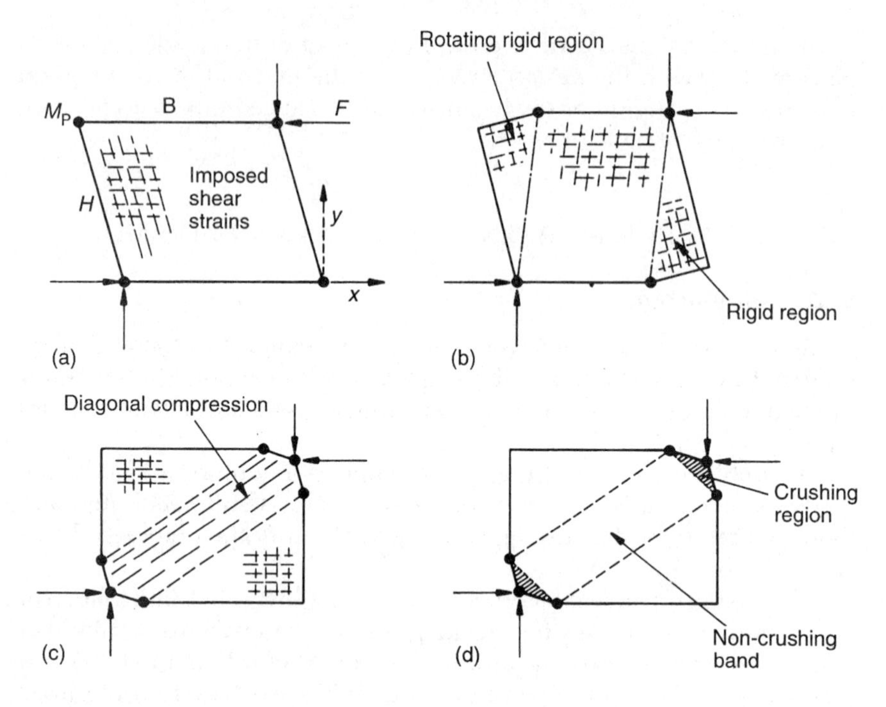

Fig. 8.11 Idealized plastic failure modes for wall frame panels: (a) shear mode S (strong frame, weak wall); (b) shear rotation mode SR (medium strength walls); (c) diagonal compression mode DC (strong wall, weak frame); (d) corner crushing mode CC (very weak frame). From Wood (1978).

A parameter m_d is introduced defined as

$$m_d = 8 M_p \gamma_m / (f_k t L^2) \tag{8.16}$$

where M_p is the lowest plastic moment of beams or columns, and f_k the characteristic strength of the masonry. This parameter which represents a frame/wall strength ratio is shown to be the factor which determined the mode of collapse.

- For $m_d < 0.25$ the collapse mode is DC (diagonal compression) or CC (corner crushing).
- For $0.25 < m_d < 1$ the collapse mode is SR (shear rotation).
- For $m_d > 1$ the collapse mode is S (shear).

(b) Design procedure

Initially the nominal value of m_d is calculated using equation (8.16) and then corrected using the factor δ_p obtained from Fig. 8.12. The corrected value (m_e) is given by $m_e = m_d / \delta_p$.

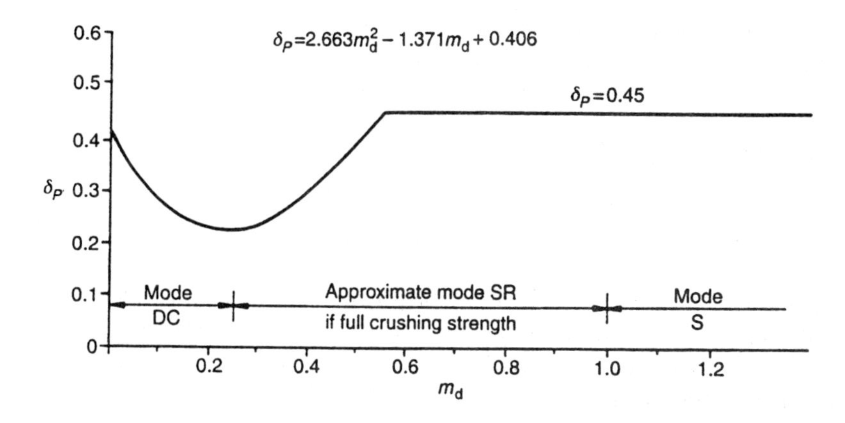

Fig. 8.12 Plot of δ_p against m_d. From Wood (1978).

Then the value of the non-dimensional parameter ϕ_s is calculated using the equation

$$\phi_s = 2/(\sqrt{m_e} + 1/\sqrt{m_e}) \tag{8.17}$$

This parameter was derived for square panels with identical beams and columns, and a correction factor Δ_ϕ must be determined for non-rectangular panels with unequal beams and columns, using Fig. 8.13 in which μ_p is defined as

$$\mu_p = \frac{\text{lowest beam plastic moment}}{\text{lowest column plastic moment}} \tag{8.18}$$

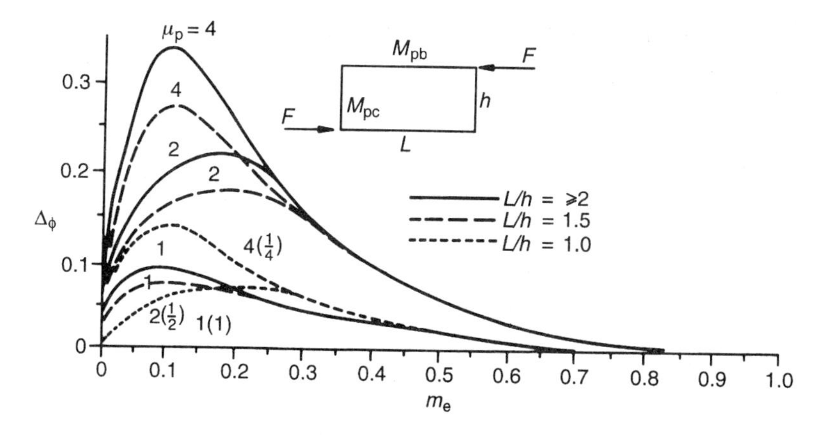

Fig. 8.13 Design chart for racking loads: optional correction Δ_ϕ added to Φ_s ($\mu = M_{pb}/M_{pc}$). From Wood (1978).

- If $\mu_p \geqslant 1$ (strong beams) use the chart directly.
- If $\mu_p < 1$ (weak beams) and $L/h = 1$ use μ_p value in brackets.
- If $\mu_p < 1$ (weak beams) and $L/h > 1$ use $\mu_p = 1$ curve.

Finally the design strength F can be determined using

$$F = (\phi_s + \Delta_\phi)\,[4\,(\text{smaller } M_p)/h + \tfrac{1}{2}\delta_p f_k tL/\gamma_m]/1.2 \tag{8.19}$$

where the factor 1.2 is an additional factor of safety introduced by Wood for design purposes and M_p is the effective plastic moment given by $Z\sigma_y/\gamma_{ms}$. For design purposes the design strength must be equal to or greater than the design load as shown in Chapter 4.

(c) Example

Assume the following dimensions and properties:

- Panel height = 2 m
- Panel length = 4 m
- Panel thickness = 110 mm
- Characteristic strength of panel = 10 N/mm^2
- Partial safety factor for masonry = 3.1
- Section modulus for each column = 600 cm^3
- Section modulus for each beam = 800 cm^3
- Yield stress of steel = 250 N/mm^2
- Partial safety factor for steel = 1.15
- Effective plastic moment for beam = $(800 \times 10^3) \times 250/(1.15 \times 10^6)$ = 174 kN/m
- Effective plastic moment for column = 130 kN/m
- $\mu_p = 1.34$
- $L/h = 2$

These give

$$m_d = \frac{8 \times 130 \times 10^6 \times 3.1}{10 \times 110 \times 4^2 \times 10^6} = 0.18$$

From Fig. 8.12, $\delta_p = 0.25$. So

$$m_e = 0.24/0.25 = 0.96$$

$$\phi_s = 2/(\sqrt{0.96} + 1/\sqrt{0.96}) = 1.0$$

From Fig. 8.13, $\Delta_\phi = 0$.

So,

$$F = 1.0\{(4 \times 130)/2 + [0.25 \times (10/3.1) \times 110 \times 4000]/2 \times 10^3\}/1.2$$
$$= 217\,(\text{frame}) + 148\,(\text{wall})$$
$$= 365\,\text{kN}$$

(d) Additional considerations

A lower limiting sliding friction wall strength F_0 is defined for the wall if composite action fails or m_d is very low:

$$F_0 = L t f_v / \gamma_{mv}$$

where

$$f_v = 0.35 + 0.6 g_A \text{ N/mm}^2 \quad \text{and} \quad f_v < 1.75 \tag{8.20}$$

for mortar designation (i), (ii) and (iii) and

$$f_v = 0.15 + 0.6 g_A \text{ N/mm}^2 \quad \text{and} \quad f_v < 1.4 \tag{8.21}$$

for mortar grade (iv) per unit area of wall cross-section due to the vertical dead and imposed load.

For the example given in section 8.2.2 (c), assuming mortar of grade (ii), f_v has a minimum value of 0.35 (for no superimposed load) and a maximum value of 1.75. Therefore taking $\gamma_{mv} = 2.5$, F_0 has a value between 62 and 308 kN depending on the value of the superimposed load on the top beam.

Design for shear in the columns and beams is based on

$$\text{column shear} = \tfrac{1}{2}(F - F_0)$$
$$\text{beam shear} = \tfrac{1}{2}(h/L)(F - F_0) \tag{8.22}$$

9

Design for accidental damage

9.1 INTRODUCTION

It would be difficult to write about the effects of accidental damage to buildings without reference to the Ronan Point collapse which occurred in 1968. The progressive collapse of a corner of a 23-storey building caused by the accidental explosion of gas which blew out the external loadbearing flank wall and the non-loadbearing face walls of one of the flats on the 18th floor made designers aware that there was a weakness in a section of their design philosophy.

The Ronan Point building was constructed of large precast concrete panels, and much of the initial concern related to structures of this type. However, it was soon realized that buildings constructed with other materials could also be susceptible to such collapse.

A great deal of research on masonry structures was therefore carried out, leading to a better understanding of the problem. Research has been undertaken in many countries, and although differences in suggested methods for dealing with abnormal loadings still exist between countries, there is also a lot of common ground, and acceptable design methods are now possible.

9.2 ACCIDENTAL LOADING

Accidental or abnormal loading can be taken to mean any loading which arises for which the structure is not normally designed. Two main cases can be identified: (1) explosive loads and (2) impact loads; but others could be added such as settlement of foundations or structural alterations without due regard to safety.

Explosions can occur externally or internally and may be due to the detonation of a bomb, the ignition of a gas, or from transportation of an explosive chemical or gas. The pressure–time curves for each of these explosive types are different, and research has been carried out to determine the exact nature of each. However, although the loading caused by an explosion is of a dynamic nature, it is general practice to assume that it is static, and design checks are normally carried out on this basis.

Accidental impact loads can arise from highway vehicles or construction equipment. A motor vehicle could collide with a wall or column of a multi-storey building or a crane load accidentally impact against a wall at any level. Both of these could cause collapse of a similar nature to those considered under explosive loading, but the method of dealing with the two types of loading may be different, as shown in section 9.4.

The risk of occurrence of an accidental load is obviously of importance in that certain risks, such as the risk of being struck by lightning, are acceptable whilst others are not. Designing for accidental damage adds to the overall cost of the building, and it is necessary to consider the degree of risk versus the increase in cost for proposed design methods to become acceptable.

The risks which society is prepared to accept can be compared numerically by considering the probability of death per person per annum for a series of types of accident. It is obvious that such estimates would vary with both time and geographical location, but values published for the United States based on accidental death statistics for the year 1966 are shown in Table 9.1.

It has also been shown that the risk for accidental damage is similar to that for fire and, since in the case of fire, design criteria are introduced, there is a similar justification for adopting criteria to deal with accidental loading. The estimates for accidental damage were based on a study of the occurrence of abnormal loadings in the United States, and Table 9.2 shows a lower bound to the number of abnormal loadings per annum.

9.3 LIKELIHOOD OF OCCURRENCE OF PROGRESSIVE COLLAPSE

Accepting that accidental loading will occur it is necessary to consider the likelihood of such loading leading to progressive collapse.

Table 9.1 Accidental death statistics for USA, 1966

Cause	*Risk, per person per annum*
Motor vehicle	2.7×10^{-4}
Falling	1.0×10^{-4}
Fire	4.0×10^{-5}
Drowning	2.8×10^{-5}
Firearms	1.3×10^{-5}
Poisoning	1.1×10^{-5}
Earthquake	8.0×10^{-7}
Lightning	5.5×10^{-7}

Table 9.2 Numbers of abnormal loadings for USA, 1966 (lower bounds)

Type	*Number, per annum*
Explosive bombing	204
Gas explosions	131
Explosion of hazardous materials	177
Highway vehicle impact	190
Total/annum	702

A range of loadbearing masonry buildings have been analysed, and basically there are three types of construction which required investigation in relation to accidental damage:

- *Case A*, where there is an outside wall without returns or only one internal return (Fig. 9.1). Removal of a panel would leave the remaining section suspended on the floor slabs above.
- *Case B*, where there is an internal wall without return (Fig. 9.2). The walls above the damaged wall will have to be carried by the floor slab.
- *Case C*, where the removal of a section of a wall imposes high local bearing stresses on a return wall or walls (Fig. 9.3). Remaining masonry is carried by return wall.

An examination of a number of both high-rise (greater than six storeys) and low-rise structures for the possible occurrence of one of the above cases, followed by the removal of a panel and analysis of the remaining structure using the yield-line theory, has shown that there would be little difficulty in designing masonry buildings to satisfy the requirements in regard to partial collapse.

In addition, experimental tests have been conducted on a section of a five-storey brickwork cross-wall structure in which sections of the main cross-walls of the ground floor were removed with a view to testing the stability of the structure in a damaged condition. The structure had not been specially designed to withstand such treatment but it remained stable throughout the tests, and it was concluded that there would be no difficulty in designing a masonry structure to provide 'alternative paths' in the case of accidental damage. In fact, in many cases there would seem to be no necessity for additional elements to secure the safety of the structure.

The likelihood of occurrence of progressive collapse in buildings similar to Ronan Point has been considered, and it is estimated that the possibility of collapse is 0.045%, i.e. 1 in every 2000 of such blocks is likely to collapse in a life of 60 years.

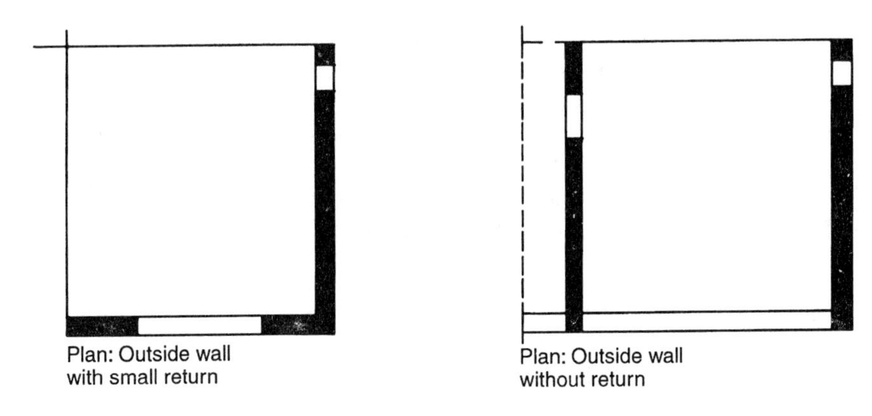

Fig. 9.1 Case A.

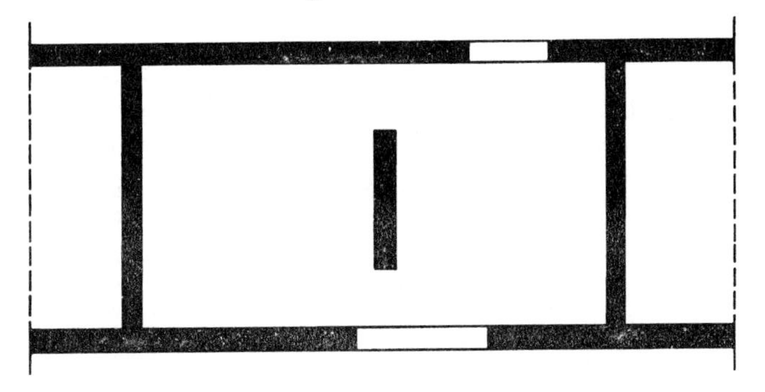

Fig. 9.2 Case B.

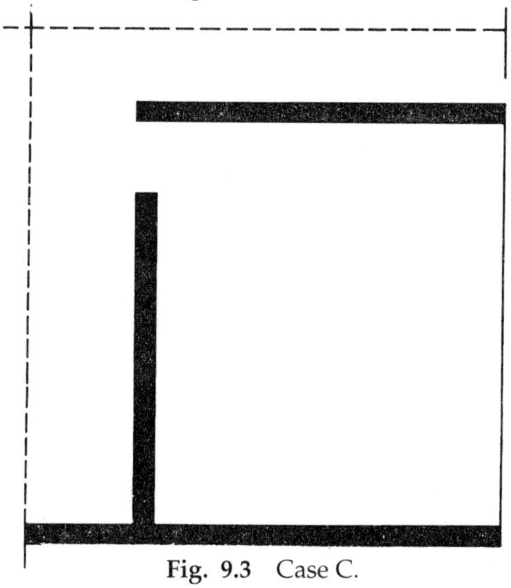

Fig. 9.3 Case C.

In summary it would appear that the risk of progressive collapse in buildings of loadbearing masonry is very small. However, against this the limited nature of the additional design precautions required to avoid such collapse are such that they add very little to the overall cost. In addition the social implications of failures of this type are great, and the collapse at Ronan Point will long be remembered. It added to the general public reaction against living in high-rise buildings.

9.4 POSSIBLE METHODS OF DESIGN

Design against progressive collapse could be introduced in two ways:

- Design against the occurrence of accidental damage.
- Allow accidental damage to occur and design against progressive collapse.

The first method would clearly be uneconomic in the general case, but it can be used to reduce the probability of local failure in certain cases. The risk of explosion, for example, could be reduced by restricting the use of gas in a building, and impact loads avoided by the design of suitable guards. However, reducing the probability does not eradicate the possibility, and progressive collapse could still occur, so that most designers favour the second approach.

The second method implies that there should be a reasonable probability that progressive collapse will not occur in the event of a local failure. Obviously, all types of failure could not be catered for, and a decision has to be made as to the extent of allowable local failure to be considered. The extent of allowable local failure in an external wall may be greater than that for an internal wall and may be related to the number of storeys. Different countries tend to follow different rules with respect to this decision.

Eurocode 6 Part 1-1 recommends a similar approach to the above but does not give a detailed example of the method of application. It refers to a requirement that there is a 'reasonable probability' that the building will not collapse catastrophically and states that this can be achieved by considering the removal of essential loadbearing members. This is essentially the same as the requirements of the British code.

Having decided that local failure may occur it is now necessary to analyse the building to determine if there is a likelihood of progressive collapse. Three methods are available:

- A three-dimensional analysis of the structure.
- Two-dimensional analyses of sections taken through the building.
- A 'storey-by-storey' approach.

The first two methods require a finite element approach and are unsuitable for design purposes, although the results obtained from such realistic methods are invaluable for producing results which can lead to meaningful design procedures. A number of papers using this approach have been published, which allow not only for the nonlinear material effects but also dynamic loading.

The third approach is conservative in that having assumed the removal of a loadbearing element in a particular storey an assessment of residual stability is made from within that storey.

These theoretical methods of analysis together with experimental studies as mentioned in section 9.3 have led to design recommendations as typified in BS 5628 (section 9.5).

9.5 USE OF TIES

Codes of practice, such as BS 5628, require the use of ties as a means of limiting accidental damage. The provisions of BS 5628 in this respect have been summarized in Chapter 4.

The British code distinguishes, in its recommendations for accidental damage design, between buildings of four storeys or less and those of five storeys or more. There are no special provisions for the first class, and there are three alternative options for the second (see Chapter 12).

It is convenient at this stage to list the types of ties used together with some of the design rules.

9.5.1 Vertical ties

These may be wall or column ties and are continuous, apart from anchoring or lapping, from foundation to roof. They should be fully anchored at each end and at each floor level.

Note that since failure of vertical ties should be limited to the storey where the accident occurred it has been suggested that vertical ties should be independent in each storey height and should be staggered rather than continuous.

In BS 5628 the value of the tie force is given as either of

$$T = (34A/8000)\,(h/t)^2\ \text{N} \tag{9.1}$$

or

$$T = 100\,\text{kN/m} \text{ length of wall or column}$$

whichever is the greater, where A = the horizontal cross-sectional area in mm^2 (excluding the non-loadbearing leaf of cavity construction but including piers), h = clear height of column or wall between restraining surfaces and t = thickness of wall or column.

The code assumes that the minimum thickness of a solid wall or one loadbearing leaf of a cavity wall is 150 mm and that the minimum characteristic compressive strength of the masonry is $5\,\mathrm{N/mm^2}$. Ties are positioned at a maximum of 5 m centres along the wall and 2.5 m maximum from an unrestrained end of any wall. There is also a maximum limit of 25 on the ratio h/t in the case of narrow masonry walls or 20 for other types of wall.

Example

Consider a cavity wall of length 5 m with an inner loadbearing leaf of thickness 170 mm and a total thickness 272 mm. Assume that the clear height between restraints is 3.0 m and that the characteristic steel strength is $250\,\mathrm{N/mm^2}$.

Using equations (9.1), tie force is the greater of

$$T = (34/8000) \times 5000 \times 170 \times (3000/272)^2 \times 10^3 = 439.5\,\mathrm{kN}$$

$$T = 100 \times 5 = 500\,\mathrm{kN}$$

Thus

$$\text{tie area} = (500/250) \times 10^3 = 2000\,\mathrm{mm^2}$$

So use seven 20 mm diameter bars. This represents a steel percentage of $(2000 \times 100)/(5000 \times 272) = 0.15\%$.

9.5.2 Horizontal ties

Horizontal ties are divided into four types and the design rules differ for each. There are (a) peripheral ties, (b) internal ties, (c) external wall ties and (d) external column ties.

The basic horizontal tie force is defined as the lesser of the two values

$$F_t = 20 + 4N_s\,\mathrm{kN}$$
$$(9.2)$$
$$F_t = 60\,\mathrm{kN}$$

where N_s = the number of storeys, but the actual value used varies with the type of tie (see below).

(a) Peripheral ties

Peripheral ties are placed within 1.2 m of the edge of the floor or roof or in the perimeter wall. The tie force in kN is given by F_t from equations (9.2), and the ties should be anchored at re-entrant corners or changes of construction.

(b) Internal ties

Internal ties are designed to span both ways and should be anchored to perimeter ties or continue as wall or column ties. In order to simplify the specification of the relevant tie force it is convenient to introduce F_t' such that

$$F_t' = F_t[(G_k + Q_k)/7.5] \times L_a/5 \text{ (kN/m width)} \quad (9.3)$$

where $(G_k + Q_k)$ is the sum of the average characteristic dead and imposed loads in kN/m^2 and L_a is the lesser of:

- the greatest distance in metres in the direction of the tie, between the centres of columns or other vertical loadbearing members, whether this distance is spanned by a single slab or by a system of beams and slabs, or
- $5 \times$ clear storey height h (Fig. 9.4).

The tie force in kN/m for internal ties is given as:

- *One-way slab* In direction of span – greater value of F_t or F_t'. Perpendicular to span – F_t.
- *Two-way slab* In both directions – greater value of F_t or F_t'.

Internal ties are placed in addition to peripheral ties and are spaced uniformly throughout the slab width or concentrated in beams with a 6 m maximum horizontal tie spacing. Within walls they are placed at a maximum of 0.5 m above or below the slab and at a 6 m maximum horizontal spacing.

(c) External wall or column ties

The tie force for both external columns and walls is taken as the lesser value of $2F_t$ or $(h/2.5)\,F_t$ where h is in metres. For columns the force is in kN whilst in walls it is kN/m length of loadbearing wall.

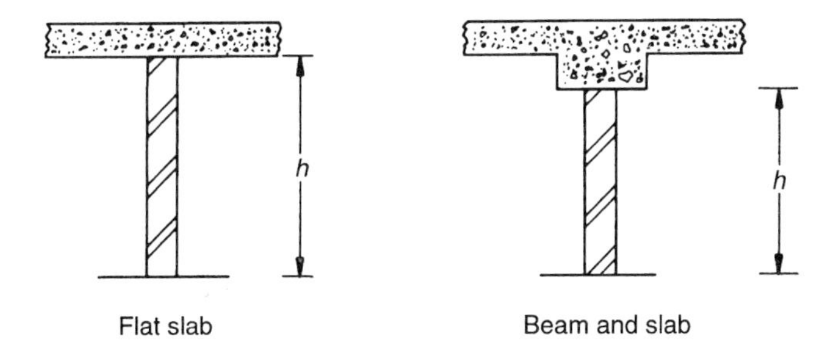

Fig. 9.4 Storey height.

Corner columns should be tied in both directions and the ties may be provided partly or wholly by the same reinforcement as perimeter and internal ties.

Wall ties should be spaced uniformly or concentrated at centres not more than 5 m apart and not more than 2.5 m from the end of the wall. They may be provided partly or wholly by the same reinforcement as perimeter and internal ties.

The tie force may be based on shear strength or friction as an alternative to steel ties (see examples).

(d) Examples

Peripheral ties

For a five-storey building

$$\text{tie force} = 20 + (5 \times 4) = 40\,\text{kN}$$

$$\text{tie area} = (40 \times 10^3)/250 = 160\,\text{mm}^2$$

Provide one 15 mm bar within 1.2 m of edge of floor.

Internal ties

Assume $G_k = 5\,\text{kN/m}^2$, $Q_k = 1.5\,\text{kN/m}^2$ and $L_a = 4\,\text{m}$. Then

$$F_t = 40\,\text{kN/m width}$$

$$F_t' = [40(5 + 1.5) \times 4]/(7.5 \times 5) = 35.5\,\text{kN/m width}$$

Therefore design for 40 kN/m both ways unless steel already provided as normal slab reinforcement.

External wall ties

Assume clear storey height = 3.0 m. Tie force is lesser of

$$2F_t = 80\,\text{kN/m length}$$

$$(h/2.5)\,F_t = (3.0/2.5) \times 40 = 48\,\text{kN/m length} \qquad \text{(which governs)}$$

Shear strength is found using Clause 25 of BS 5628,

$$f_v = 0.35 + 0.6g_A \quad (\text{max. } 1.75)$$

or

$$f_v = 0.15 + 0.6g_A \quad (\text{max. } 1.4)$$

depending on mortar strength. From Clause 27.4 of BS 5628,

$$\gamma_{mv} = 1.25$$

Assume mortar to be grade (i).

Taking g_A, the design vertical load per unit area due to dead and imposed load, as zero, is conservative and equivalent to considering shear strength due to adhesion only. That is design shear strength on each surface $= f_v/\gamma_{mv} = 0.35/1.25 = 0.28\,\text{N/mm}^2$.

Combined resistance in shear on both surfaces is

$$2 \times \text{shear stress} \times \text{area} = 2 \times 0.28 \times (110 \times 1000/1000) = 61.6\,\text{kN/m}$$

In this example the required tie force of 48 kN/m is provided by the shear resistance of 61.6 kN/m, and additional steel ties are not required. If the shear resistance had been less than the required tie force, then the steel provided would be based on the full 48 kN/m.

Alternatively the required resistance may be provided by the frictional resistance at the contact surfaces (Fig. 9.5). This calculation requires a knowledge of the dead loads from the floors and walls above the section being considered.

Assume dead loads as shown in Fig. 9.6. Using a coefficient of friction of 0.6 the total frictional resistance on surfaces A and B is

$$(20 + 10)0.6 + (20 + 10 + 18)\,0.6 = 46.8\ \text{kN/m}$$

which would be insufficient to provide the required tie force. Note that the code states that the calculation is based on shear strength or friction (but not both).

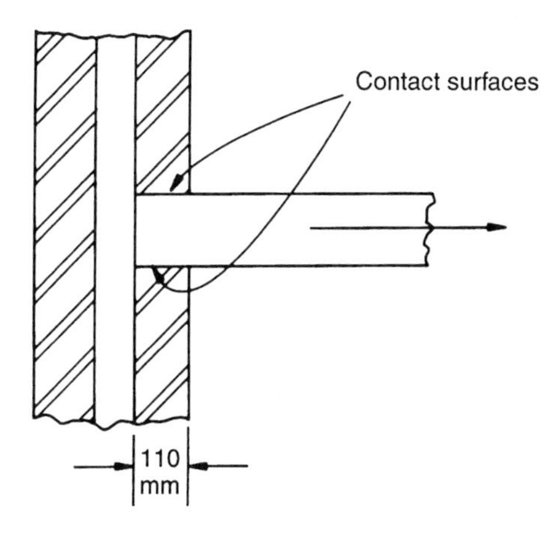

Fig. 9.5 Surfaces providing frictional resistances.

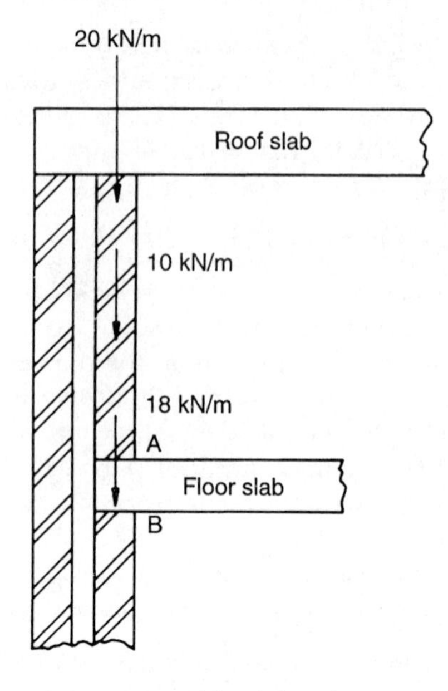

Fig. 9.6 Dead load distribution.

10

Reinforced masonry

10.1 INTRODUCTION

Possible methods of construction in reinforced masonry (illustrated in Fig. 10.1) may be summarized as follows:

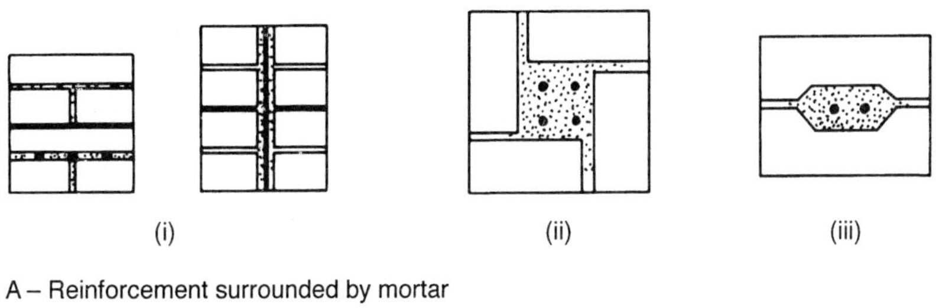

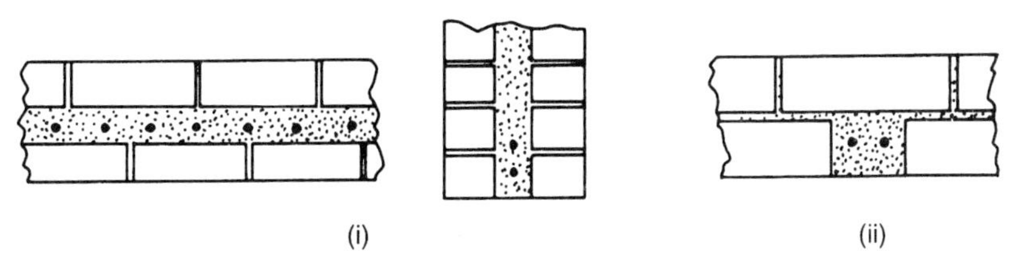

Fig. 10.1 Methods of reinforcing brickwork.

(A) Reinforcement surrounded by mortar
- (i) in bed joints or collar joints
- (ii) in pockets formed by the bond pattern of units
- (iii) in pockets formed by special units

(B) Reinforcement surrounded by concrete
- (i) in cavity between masonry leaves

(ii) in pockets formed in the masonry
(iii) in the cores of hollow blocks
(iv) in U-shaped lintel units

Type A(i) is suitable for lightly reinforced walls when the steel is placed in the bed joints, for example to improve the resistance of a wall to lateral loading. Larger-diameter bars or reinforcement in two directions can be accommodated when the steel is placed in the collar joint of a stretcher bond wall. Such an arrangement is suitable for a shear wall. Type A(ii) includes Quetta bond and may be used as a means of introducing steel for controlling earthquake or accidental damage. The use of specially shaped units produces a similar result. In these methods the steel is placed and surrounded by mortar as the work proceeds.

In types B(i) and (ii) the spaces for the reinforcing bars are larger and are filled with small aggregate concrete. Types B(iii) and (iv) are used for reinforced concrete blockwork, vertically and/or horizontally reinforced. In this case, the cavity pockets or cores may be filled as the masonry is laid in lifts up to 450 mm in height or, alternatively, walls may be built up to 3 m height before placing the infill concrete. In the latter case, provision has to be made for cleaning debris from the internal spaces before filling with concrete. This technique is suitable for walls, beams and columns and can accommodate any practicable amount of reinforcement. In particular, grouted cavity beams can be reinforced with vertical and diagonal bars for shear resistance.

10.2 FLEXURAL STRENGTH

10.2.1 Stress–strain relationships

In order to develop design equations for elements subject to bending it is necessary to assume ideal stress–strain relationships for both the masonry and the reinforcement.

As far as the masonry is concerned the approximate parabolic distribution shown in Fig. 3.5 may be further simplified to a rectangular distribution in which the stress is assumed to be constant and equal to f_k/γ_{mm} (Fig. 10.2).

As far as the steel is concerned the relationship is assumed to be as shown in Fig. 10.3 where f_y, the characteristic tensile strength of the reinforcement, is assumed to be 250 N/mm^2 for hot-rolled deformed high-yield steel and 460 N/mm^2 for hot-rolled plain, cold-worked steel and stainless-steel bars.

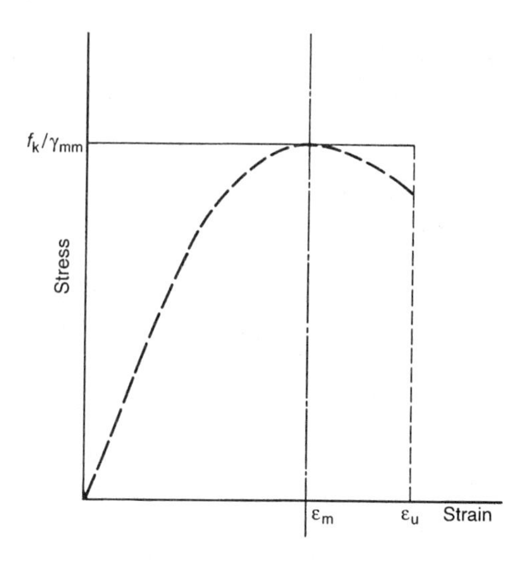

Fig. 10.2 Idealized stress–strain relationship for brickwork.

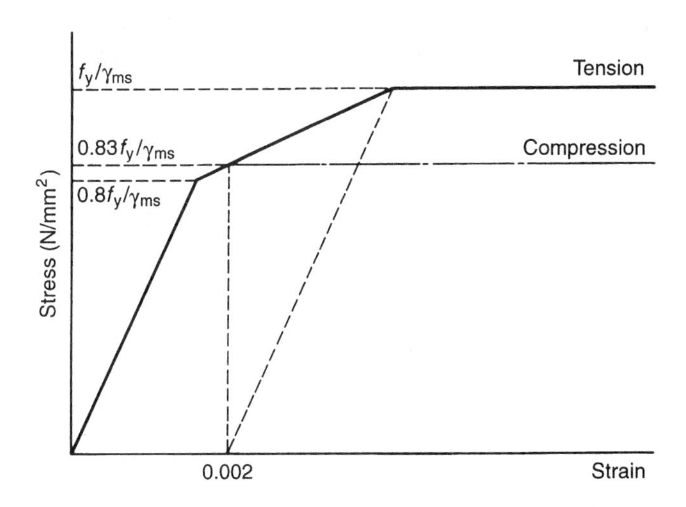

Fig. 10.3 Idealized stress–strain relationship for reinforcement (BS 5628).

10.2.2 Additional assumptions and limitations

In addition to the idealization of the stress–strain relationships further assumptions are introduced as follows:

1. Plane sections remain plane after bending.
2. The tensile strength of the masonry is ignored.
3. The effective span of simply supported or continuous members is taken as the smaller of (i) the distance between support centres and (ii) the clear distance between supports plus the effective depth.
4. The effective span of cantilevers is taken as the smaller of (i) the distance between the end of the cantilever and the centre of its support and (ii) the distance between the end of the cantilever and the face of the support plus half its effective depth.
5. The ratio of span to effective depth is not less than 1.5 otherwise the beam would have to be designed as a deep beam and the basic equations would not be applicable.
6. The strains in both materials are directly proportional to the distances from the neutral axis.
7. The section is under-reinforced so that the strain in the reinforcement reaches the yield value ε_y whilst the maximum strain in the masonry is still below the ultimate value ε_u. (A limiting strain distribution can be defined in which the reinforcement is at ε_y and the masonry at ε_u (Fig. 10.4).)
8. Although design is based on the ultimate limit state, recommendations are included in the codes of practice to ensure that the serviceability states of deflection and cracking are not reached. These recommendations are given as limiting ratios of span to effective depth. (See Tables 8 and 9 of BS 5628: Part 2, and similar recommendations in EC6 Part 1-1.)
9. To ensure lateral stability beams should be proportioned so that (i) for simply supported or continuous beams the distance between lateral restraints does not exceed the lesser of $60\,b_c$ and $250b_c^2/d$, and (ii) for

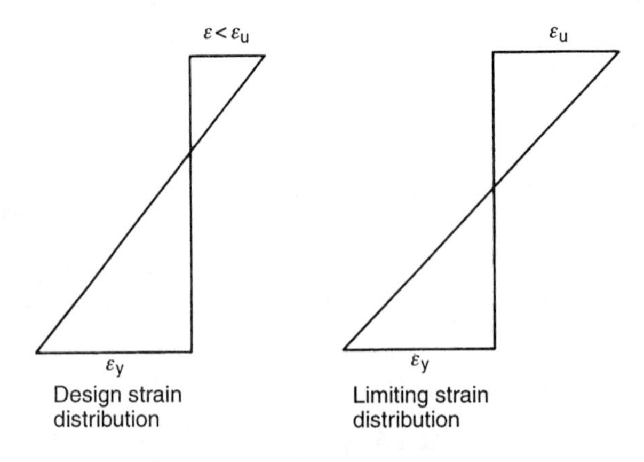

Fig. 10.4 Strain distributions.

cantilevers the distance between the end and the support does not exceed the lesser of $25b_c$ and $100b_c^2/d$, where b_c is the width of the compression face midway between restraints and d is the effective depth.

10.2.3 Design equations

Considering a rectangular cross-section subjected to bending and using the assumptions listed above the basic equations required for design can be derived as follows.

In Fig. 10.5 the strain distribution shows that the steel has reached yield strain and the maximum masonry strain is less than the ultimate value (assumption 7). Also the stress in the compressive zone is constant at f_k/γ_{mm} (stress–strain relationship for masonry).

Taking moments about the centroid of the compression block gives the design moment of resistance M_d

$$M_d = A_s z f_y/\gamma_{ms} \tag{10.1}$$

where

$$z = (d - d_c/2) \tag{10.2}$$

Equating the total tensile force to the total compressive force gives

$$A_s f_y/\gamma_{ms} = bd_c f_k/\gamma_{mm}$$

so that

$$d_c/d = A_s f_y \gamma_{mm}/bdf_k\gamma_{ms} \tag{10.3}$$

Substituting (10.3) into (10.2) gives

$$z = d(1 - 0.5A_s f_y \gamma_{mm}/bdf_k\gamma_{ms}) \tag{10.4}$$

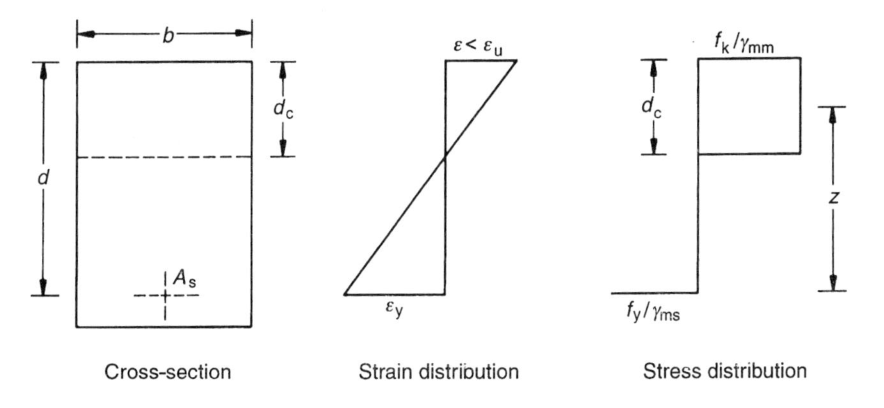

Fig. 10.5 Strain and stress distribution in section.

The design of sections for bending only can be carried out using equations (10.1) and (10.4) although it would be necessary to solve a quadratic equation in A_s to determine the area of reinforcement. This is considered to be inconvenient and the British code includes tables and charts for the direct solution. An alternative method is shown in section 10.2.4.

The assumption of a limiting strain distribution as shown in Fig. 10.4 imposes an upper bound to the value of M_d. Theoretically this limit can be determined from the ratio

$$d_c/d = \varepsilon_u/(\varepsilon_u + \varepsilon_y)$$

which is dependent on the maximum strain ε_u (taken as 0.0035 in the code) and ε_y (which is dependent on the type of steel). It can be shown that the theoretical limiting value of M_d/bd^2 for the assumed stress–strain distribution is given approximately by

$$M_d/bd^2 = 0.4 f_k/\gamma_{mm} \tag{10.5}$$

Adoption of this limit precludes brittle failure of the beam.

10.2.4 Design aid

Equations (10.1), (10.4) and (10.5) can be represented graphically, as shown in Figs 10.6 and 10.7 for particular values of f_k, γ_{mm} and γ_{ms}. The graphs relate the three parameters M_d/bd^2, f_k and ρ so that given any two the third can be determined directly from the graph. The steel ratio ρ is equal to A_s/bd.

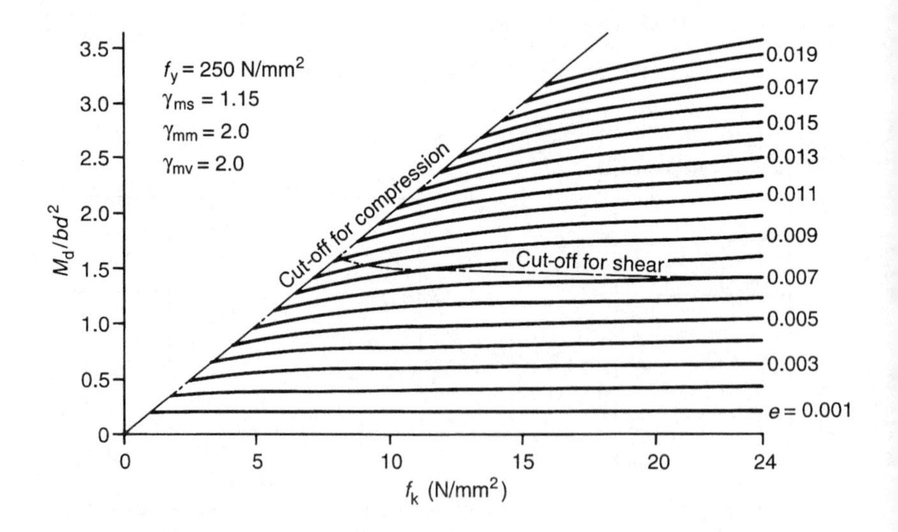

Fig. 10.6 Design aid for bending ($f_y = 250$ N/mm^2).

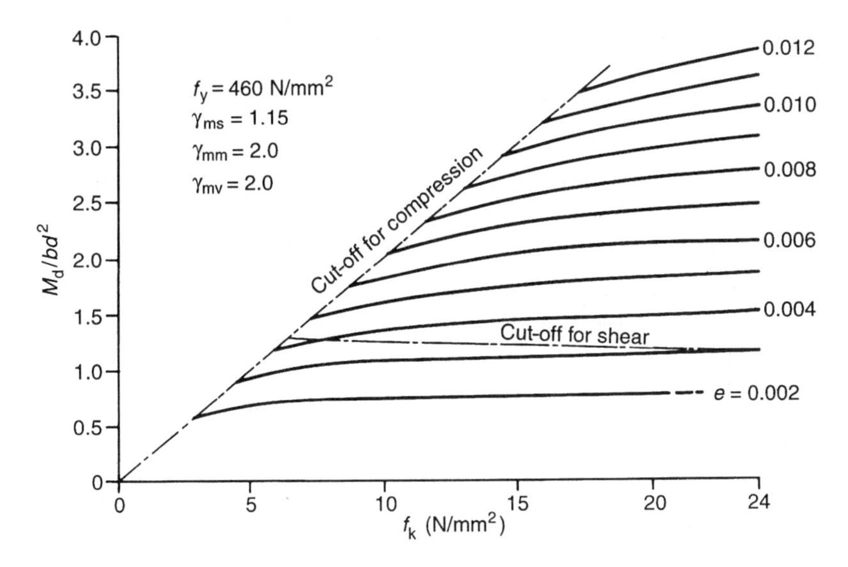

Fig. 10.7 Design aid for bending ($f_y = 460\ \text{N/mm}^2$).

Since values of M_d/bd^2 are approximately constant for a particular value of ρ this shows that the characteristic strength of the masonry has limited influence on the design.

10.2.5 Example

Design a simply supported brickwork beam of span 4 m and of section 215 mm × 365 mm to carry a moment of 24 kN m assuming that the characteristic strength of the material is 19.2 N/mm². Assume also that $\gamma_{mm} = 2.0$ and $f_y = 250\ \text{N/mm}^2$.

The effective depth of the reinforcement allowing for 20 mm diameter bars and a cover of 20 mm would be 365 − 20 − 10 = 335 mm. So

$$M_d/bd^2 = (24 \times 10)/(215 \times 335) = 0.99$$

Using Fig. 10.6 with $f_k = 19.2\ \text{N/mm}^2$ gives

$$A_s/bd = 0.005$$

$$A_s = 0.005 \times 215 \times 335 = 360\ \text{mm}^2$$

Use two 16 mm diameter bars providing 402 mm².

Check for stability. The lesser of $60b$ and $250b_c^2/d$ is $60 \times 215 = 12.9$ m. This is greater than 4 m and therefore acceptable.

Note that since the intersection of the lines for $M_d/bd^2 = 0.99$ and $f_k = 19.2\,\text{N/mm}^2$ in Fig. 10.6 is below the cut-off line for shear, the design will be safe in shear.

10.3 SHEAR STRENGTH OF REINFORCED MASONRY

10.3.1 Shear strength of reinforced masonry beams

As in reinforced concrete beams, shear transmission across a crack in a reinforced masonry beam can take place by one or more of the following mechanisms:

- Compression zone transmission resulting from the shear resistance of the masonry.
- 'Aggregate interlock' by frictional forces across the crack.
- 'Dowel effect' from the shear force developed by the reinforcing bars crossing the crack.

The relative importance of these effects depends on the construction of the beam. Thus in a masonry cross-section, a shear crack develops stepwise through the mortar joints and therefore aggregate interlock will be limited. Also in a beam of this type, where the reinforcement is placed in the lowermost bed joint of a brick masonry beam, dowel effect will be restricted by the low tensile strength of the brick–mortar joint. In a grouted cavity beam on the other hand, both aggregate interlock and dowel action will be developed in the concrete core and thus the overall shear strength of the beam will be greater than in a brick masonry section of the same overall size. In concrete blockwork it is usual to employ a U-shaped lintel block in the lowermost course which will result in greater shear resistance from dowel effect.

Shear resistance of a reinforced masonry beam is also influenced by the shear span ratio of the beam. In the simplest case of a simply supported beam loaded by two equal symmetrically placed loads, this ratio is defined by the parameter a/d, where a is the distance of the load from the support and d is the effective depth. As the shear span ratio is reduced below about 6 the shear strength increases quite rapidly, as shown in Fig. 10.8. The explanation for this is that when the shear span ratio is low, the beam behaves after the manner of a tied arch, as suggested in Fig. 10.9.

In reinforced concrete beams, shear strength increases with increase in the steel ratio. As might be expected, this is also the case in grouted cavity reinforced masonry beams. However, brick masonry beams do not show such an increase, no doubt because dowel effect is not developed.

Shear reinforcement in the form of vertical steel or bent-up bars can be introduced in grouted cavity beams but the scope for such reinforcement of masonry sections is limited.

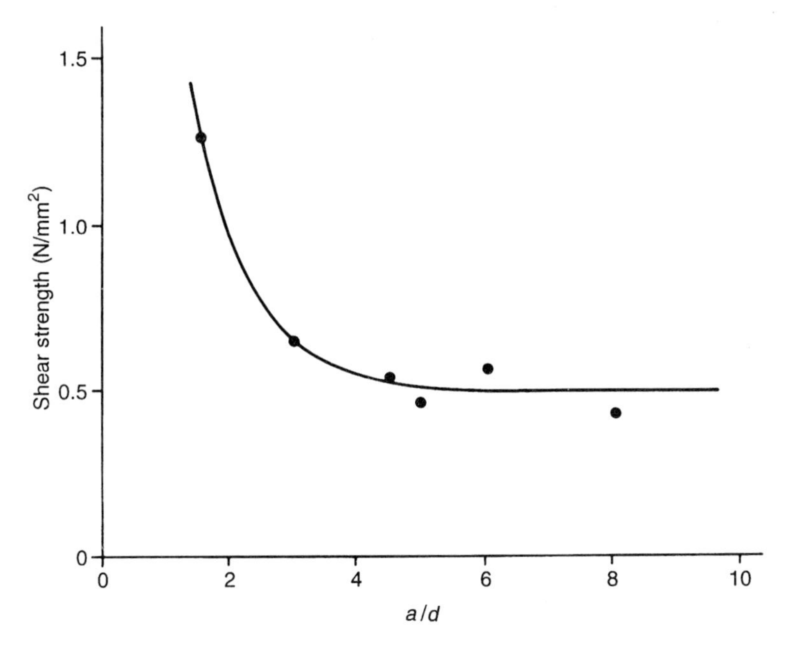

Fig. 10.8 Shear strength versus shear span ratio for grouted cavity brickwork beam.

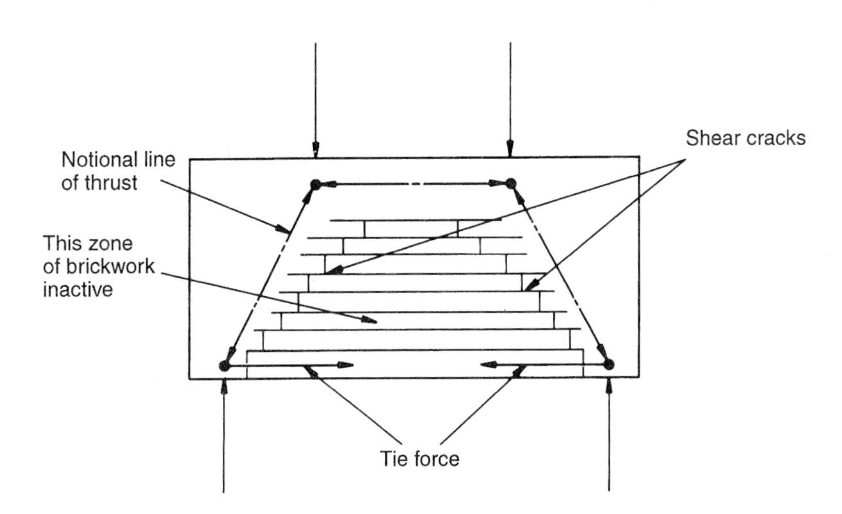

Fig. 10.9 Internal tied arch action in a reinforced brickwork beam having low shear span ratio.

10.3.2 Shear strength of rectangular section reinforced masonry beams

The method of calculating the flexural strength of reinforced masonry beams is discussed in section 10.2. It is also necessary to ensure that the shear stress in a beam does not exceed the design shear strength of the material, i.e.

$$V/bd > f_v/\gamma_{mv} \tag{10.6}$$

where V is the design shear force at a section, b and d are respectively the breadth and effective depth, f_v the characteristic shear strength and γ_{mv} the partial safety factor for shear.

As an illustration of the influence of shear strength on the design of rectangular section beams, it is possible to plot a 'cut-off' line on Figs. 10.6 and 10.7 defining the M_d/bd^2 value above which shear will be the limiting factor. This has been derived by assuming that the shear span is $a = M_{max}/V$, so that, referring to equation (10.6):

$$M_{max}/(a/d)\,bd^2 > f_v/\gamma_{mv}$$

or

$$M_{max}/bd^2 > (f_v/\gamma_{mv})\,(a/d)$$

In Figs 10.6 and 10.7, $f_v = 0.35(1+17.5\rho)$, $a/d = 6$ and $\gamma_{mv} = 2.0$. For these conditions it is apparent that shear strength will be a limiting factor for steel ratios above 0.007–0.009 and 0.003–0.004 for $f_y = 250$ N/mm^2 and 460 N/mm^2, respectively unless shear reinforcement is provided.

The provision of shear reinforcement presents no difficulty in grouted cavity sections. It is possible in brick masonry sections by incorporating pockets in the masonry after the manner of Quetta bond and in some types of hollow concrete blockwork. BS 5628: Part 2 gives the following formula for the spacing of shear reinforcement where it is required:

$$A_{sv}/s_v \geqslant b[v - (f_v/\gamma_{mv})]\gamma_{ms}/f_y \tag{10.7}$$

where A_{sv} is the cross-sectional area of reinforcing steel resisting shear forces, b is the width of the section, f_v is the characteristic shear strength of masonry, f_y is the characteristic tensile strength of the reinforcing steel, s_v is the spacing of shear reinforcement along the member, but not to exceed $0.75d$, v is the shear stress due to design loads but not to exceed $2.0/\gamma_{mv}$ N/mm^2, γ_{ms} is the partial safety factor for the strength of steel and γ_{mv} is the partial safety factor for shear strength of masonry.

10.3.3 Resistance to racking shear

Shear walls are designed to resist horizontal forces in their own plane. In certain cases flexural stresses are significant and the strength of the wall may be closely predicted by assuming that all the vertical reinforcement has yielded and that the compression zone is located at the 'leeward' toe of the wall. If, however, flexural stresses are reduced by the presence of vertical loading it has been found that a lower bound on shear strength of 0.7 N/mm^2 may be assumed for walls having more than 0.2% reinforcement. If the vertical compression is higher than 1.0 N/mm^2 this will be exceeded by the strength of an unreinforced wall and in such a case the effect of the reinforcement could be neglected in assessing the design strength. The presence of reinforcement, however, is important in seismic conditions in developing a degree of ductility and in limiting damage.

10.4 DEFLECTION OF REINFORCED MASONRY BEAMS

The deflection of a reinforced masonry beam can be calculated in a similar way to that of a reinforced concrete beam with suitable adjustments for different material properties. Experiment has shown that the following moment–curvature relationship can be assumed:

$$\theta = \frac{M}{EI_u} + \frac{M - M_{cr}}{0.85EI_{cr}} \qquad (10.8)$$

where M is the applied moment, EI_u is the flexural rigidity of the transformed uncracked section, EI_{cr} is the flexural rigidity of the transformed cracked section, $M_{cr} = I_{cr} f_t / (H - d_c)$ is the cracking moment, f_t is the apparent flexural tensile strength of the masonry (or composite brick/concrete in a grouted cavity beam), H is the overall depth of the section and d_c is the neutral axis depth.

The mid-span deflection of a beam of span L for various loading cases is given in Table 10.1 in terms of θ.

Table 10.1 Relationship between curvature and deflection at mid-span for various loading cases

Loading	*Central deflection*
Concentrated load at mid-span	$\theta L^2/12$
Uniformly distributed load	$\theta L^2/9.6$
Equal end moments	$\theta L^2/8$

10.5 REINFORCED MASONRY COLUMNS, USING BS 5628: PART 2

10.5.1 Introduction

Elements such as columns, which are subjected to both vertical loading and bending, are classified as either short or slender and different equations are used for the design of the two classes. Additionally bending may be about one or two axes so that a number of cases can be identified.

In the code, short columns are defined as those with a slenderness ratio (see Chapter 5) of less than 12 and, although uniaxial and biaxial bending are discussed for short columns, very little guidance is given for the case of biaxial bending of slender columns.

The stress–strain relationships assumed for the masonry and the reinforcement are the same as those assumed for the case of bending only and are as described in section 10.2.1.

10.5.2 Additional assumptions and limitations

Assumptions 1, 2 and 6 given in section 10.2.2 are assumed to apply also to column design. Additionally:

- The effective height and thickness are as given in Chapter 5.
- The maximum strain in the outermost compression fibre at failure is taken as 0.0035.

This latter assumption together with the assumption that the strains in both materials are directly proportional to the distances from the neutral axis are used as the starting point for considering a number of possible cases (see Fig. 10.10).

For each case the maximum compressive strain is assumed to be 0.0035 and the maximum compressive stress f_k/γ_{mm}.

Also for each case the strain at the level of the reinforcement near the more highly compressed face (ε_1) is of such a magnitude that the stress at this level (f_{s1}) is equal to $0.83f_k$.

The strain (ε_2) at the level of the reinforcement near the least compressed face is a function of d_c, the depth of the masonry in compression. In practice the value of d_c is assumed and the stress in this reinforcement (f_{s2}) determined by means of the following simplifying assumptions.

1. The value of d_c is assumed to be greater than $2d_1$.
2. If d_c is chosen to be between $2d_1$ and $t/2$ then f_{s2} is taken as f_y.
3. If d_c is chosen to be between $t/2$ and $(t-d_2)$ then f_{s2} is found by interpolation using

$$f_{s2} = 2f_y(t-d_2-d_c)/(t-2d_2)$$

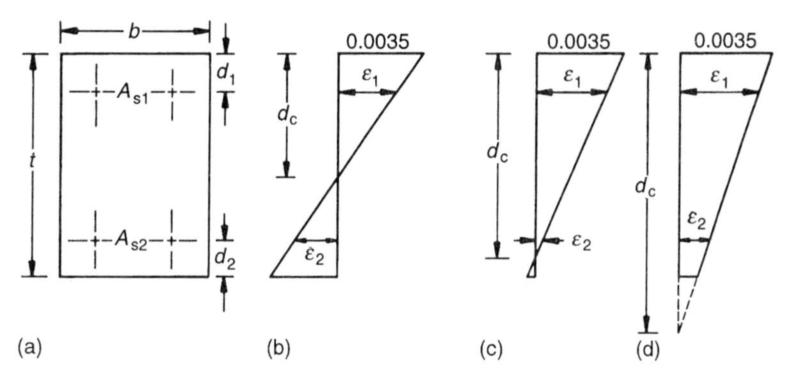

Fig. 10.10 Strain distribution in columns.

4. If d_c is chosen to be between $(t - d_2)$ and t then $f_{s2} = 0$.
5. For a strain distribution similar to Fig. 10.10(d) the stress in A_{s2} will be compressive and will vary between 0 and $-0.83f_y$. An additional assumption regarding the value of the strain at depth t would be required in order to determine the interpolated value.

10.5.3 Short columns

(a) Uniaxial bending

Based on the assumption given above three cases for the design of short columns subjected to bending about one axis are outlined in the code.

Case (a)

This case applies when the design axial load, N, is less than the value of the design axial load resistance, N_d, given by

$$N_d = f_k b(t - 2e_x)/\gamma_{mm} \tag{10.9}$$

For this case only a minimum amount of reinforcement is required and the code suggests that designers should consider if design in accordance with BS 5628: Part 1 would be more appropriate.

Case (b)

This case applies when the design axial load N is greater than the value of the design axial load resistance N_d, given in case (a). The basic equations can be derived in a similar manner to the method used in section 10.2.3 for bending by

- determining the total force in the stress diagram and
- taking moments about the mid-section.

The resulting equations are:

$$N_d = \frac{f_k b d_c}{\gamma_{mm}} + \frac{0.83 f_y A_{s1}}{\gamma_{ms}} - \frac{f_k A_{s2}}{\gamma_{ms}} \qquad (10.10)$$

$$M_d = \frac{0.5 f_k b d_c (t - d_c)}{\gamma_{mm}} + \frac{0.83 f_y A_{s1}(0.5t - d_1)}{\gamma_{ms}} + \frac{f_k A_{s2}(0.5t - d_2)}{\gamma_{ms}} \qquad (10.11)$$

The values of N_d and M_d calculated using these equations must be greater than N and M, the applied axial load and bending moment. Trial sections and areas of reinforcement are first assumed and then f_{s2} determined from an assumed value of d_c following the method outlined in section 10.5.2. This method is cumbersome and interaction diagrams are available for a more direct solution of the equations. In these diagrams $M/bt^2 f_k$ is plotted against $N/bt^2 f_k$ for a range of values of ρ/f_k and separate diagrams are available for different values of the ratio d/t and f_y.

Case (c)
In this case, which is used when the eccentricity M/N is greater than $(t/2 - d_1)$, the axial load is ignored and the section designed to resist an increased moment given by

$$M_a = M + N(t/2 - d_1) \qquad (10.12)$$

For this method the area of tension reinforcement can be reduced by $N\gamma_{ms}/f_y$.

(b) Biaxial bending

For short columns the code states that it is usually sufficient to design for uniaxial bending even when significant moments occur about both axes. However, a method is included to deal with the biaxial case by increasing the moment about one of the axes in accordance with

$$M'_x = M_x + \alpha\left(\frac{p}{q}\right) \quad \text{for } \frac{M_x}{p} > \frac{M_y}{q} \qquad (10.13)$$

$$M'_y = M_y + \alpha\left(\frac{q}{p}\right) \quad \text{for } \frac{M_x}{p} < \frac{M_y}{q} \qquad (10.14)$$

Taking the design axial load resistance for the complete section (A_m) and ignoring all bending as

$$N_{dz} = f_k A_m \qquad (10.\ 15)$$

the value of α can be determined from Table 10.2.

Table 10.2 Values of α for biaxial bending of a short column

N/N_{dz}	α
0	1.00
0.1	0.88
0.2	0.77
0.3	0.65
0.4	0.53
0.5	0.42
>0.6	0.30

Note that Table 10.2 and the above equations are different from those given in the code as originally published.

10.5.4 Slender columns

Columns which have a slenderness ratio between 12 and 27 are considered to be slender. Such columns might have an appreciable horizontal deflection due to the vertical load (Fig. 10.11) and this can be allowed for in design by increasing the eccentricity.

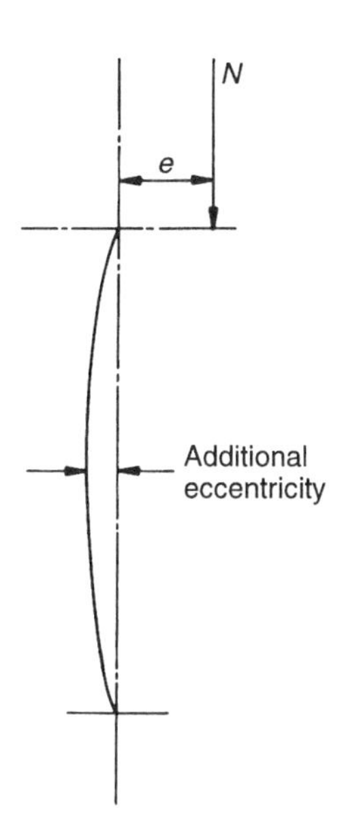

Fig. 10.11 Additional eccentricity for slender columns.

The moment due to this additional eccentricity is given by the equation

$$M_a = N(h_{ef})^2/(2000t) \quad (10.16)$$

For uniaxial bending slender columns can be designed using the method outlined in section 10.5.3 but allowing for the additional moment M_a given above.

As stated in section 10.5.1 very little guidance is given in the code for the design of slender columns subjected to biaxial bending although it states that it is essential to take account of such cases. Design can be carried out using similar methods to those used for reinforced concrete columns but applying the assumptions given in sections 10.5.1 and 10.5.2.

10.5.5 Example

A brickwork column of section 460 mm × 460 mm is to carry an axial load of 800 kN and a moment of 50 kN m. Assuming that the reinforcement is placed such that $d_2 = d_1 = 130$ mm design the column for (1) an effective height of 4.5 m and (2) an effective height of 6.0 m. Take $f_k = 13\,\text{N/mm}^2$, $f_y = 460\,\text{N/mm}^2$, $\gamma_{mm} = 2.3$.

Case 1

In this case

$$\text{slenderness ratio} = 4.5/0.46 = 9.8 \quad \text{i.e. short column}$$

$$\text{resultant eccentricity} = 800 = 0.0625\,\text{m}$$

Using equation (10.9)

$$\begin{aligned} N_d &= f_k b(t - 2e_x)/\gamma_{mm} \\ &= 13 \times 460(460 - 125)/2.3 \\ &= 871\,000 = 871\,\text{kN} \end{aligned}$$

Since $N < N_d$, case (a) applies and only a minimum amount of reinforcement is required.

Design in accordance with BS 5628: Part 1 might be more appropriate.

Case 2

In this case

$$\text{slenderness ratio} = 6.0/0.46 = 13 \quad \text{i.e. slender column}$$

$$\begin{aligned} \text{additional moment } M_a &= N(h_{ef})^2/(2000t) \\ &= 800 \times 6^2/(2000 \times 0.46) \\ &= 31.3\,\text{kN m} \end{aligned}$$

Design for axial load = 800 kN and

$$\text{moment} = 50 + 31.3 = 81.3\,\text{kN m}$$

Assume that $d_c = 300$ mm and $A_{s1} = A_{s2} = 905\,\text{mm}^2$ (two T24 bars). Since d_c is between $t/2$ and $(t-d_2)$, f_{s2} can be determined from

$$f_{s2} = 2f_y\left(\frac{t-d_2-d_c}{t-2d_2}\right) = 2f_y\left(\frac{460-130-300}{460-260}\right)$$

$$= 0.3f_y$$

Take $f_{s1} = 0.83f_y$. Then

$$N_d = \frac{f_k b d_c}{\gamma_{mm}} + \frac{f_y A_{s1}}{\gamma_{ms}}(0.83-0.3)$$

$$= \frac{13 \times 460 \times 300}{2.3 \times 10^3} + \frac{460 \times 905 \times 0.53}{1.15 \times 10^3}$$

$$= 780 + 192$$

$$= 972 \quad \text{which is adequate}$$

$$M_d = \frac{0.5 f_k b d_c (t-d_c)}{\gamma_{mm}} + \frac{f_y A_{s1}(0.5t-d_1)(0.83+0.3)}{\gamma_{ms}}$$

$$= \frac{0.5 \times 13 \times 460 \times 300 \times 160}{2.3 \times 10^6} + \frac{460 \times 905 \times 100 \times 1.13}{1.15 \times 10^6}$$

$$= 62.4 + 40.9$$

$$= 103.3\,\text{kN m} \quad \text{which is adequate.}$$

10.6 REINFORCED MASONRY COLUMNS, USING ENV 1996-1-1

10.6.1 Introduction

The Eurocode does not refer separately to specific design procedures for reinforced masonry columns although in section 4.7.1.6 of the code reference is made to reinforced masonry members subjected to bending and/or axial load. In the section a diagram showing a range of strain distributions, in the ultimate state, for all the possible load combinations is given and these are based on three limiting strain conditions for the materials.

1. The tensile strain of the reinforcement is limited to 0.01.
2. The compressive strain in the masonry due to bending is limited to -0.0035.

3. The compressive strain in the masonry due to pure compression is limited to -0.002.

Using these conditions a number of strain profiles can be drawn.

For example if it is decided that at the ultimate state the strain in the reinforcement has reached its limiting value then the range of strain diagrams take the form shown in Fig. 10.12. In Fig. 10.12 the strain diagrams all pivot about the point A, the ultimate strain in the reinforcement. Line 2 would represent the strain distribution if the ultimate compressive strain was attained in the masonry at the same time as the ultimate strain was reached in the reinforcement and line 1 an intermediate stage. In the Eurocode additional strain lines, such as line 3, are included in the diagram but since no tension is allowed in the masonry these strain distributions would require upper reinforcement.

If the limiting condition is assumed to be that the strain in the masonry has reached its limiting value then the strain distribution diagrams would be as shown in Fig. 10.13. In Fig. 10.13 the strain diagrams all pivot about the point B, the ultimate compressive strain in the masonry. Line 3 would represent the strain distribution if the ultimate tensile strain was attained in the reinforcement at the same time as the ultimate compressive strain was reached in the masonry and line 2 an intermediate stage. Line 1, representing the limiting line for this range, occurs when the depth of the compression block equals the depth of the section. Compare section 10.5.2.

To allow for pure compression, with a limiting strain value of -0.002, the Eurocode allows for a third type of strain distribution as shown in Fig. 10.14. In Fig. 10.14 the strain diagrams all pivot about the point C at

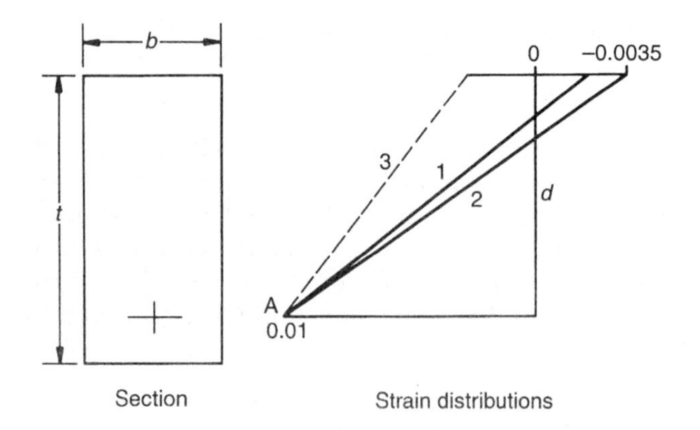

Fig. 10.12 Strain diagrams with reinforcement at ultimate.

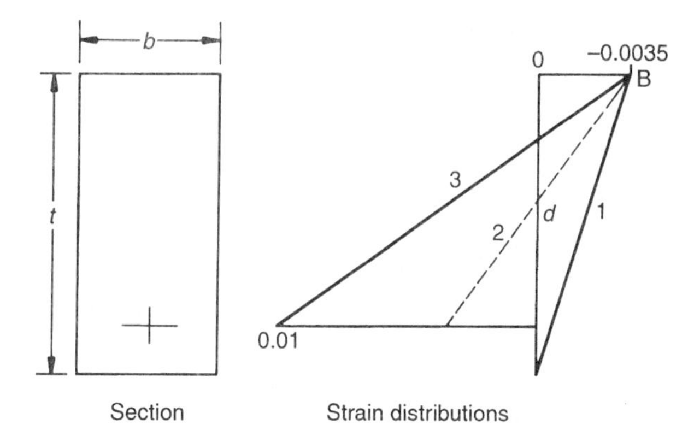

Fig. 10.13 Strain diagrams with masonry at ultimate.

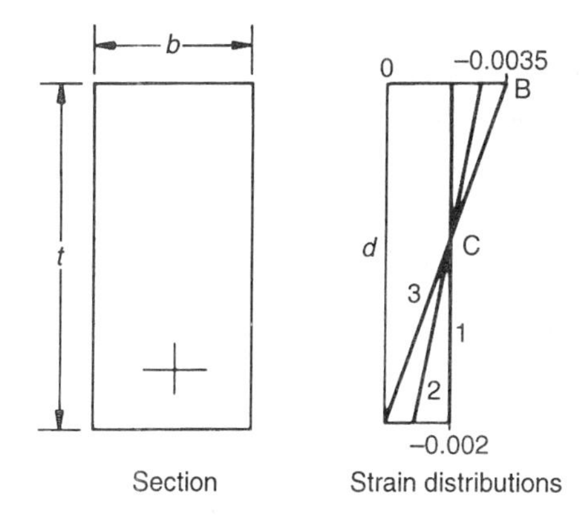

Fig. 10.14 Strain diagrams with pure compression limited to a strain of -0.002.

a level of $3t/7$ from the most compressed face. Line 3 would represent the strain distribution if the ultimate compressive strain was attained in the upper face of the masonry together with no strain in the lower face and line 2 an intermediate stage. Line 1 would represent the strain distribution for pure compression.

All the strain diagrams represented in Figs. 10.12 to 10.14 are combined into a single diagram in the Eurocode to cover various combinations of bending and/or axial loading. For reinforced masonry columns subjected to bending and compression the strain diagrams of Fig. 10.12 would be excluded to avoid the possibility of brittle failure. The strain diagrams shown in Fig. 10.13 are similar to those shown in Fig. 10.10 (a) and (b) using BS 5628.

10.6.2 Comparison between the methods of BS 5628 and ENV 1996-1-1

(a) Strain diagrams

The strain diagrams shown in Fig. 10.14 differ from those used in BS 5628 in the selection of the pivotal point; the Eurocode uses the pivot C whilst BS 5628 uses the pivot B. As a result of this, Eurocode calculations in this range might result in the maximum compressive stress in the masonry being less than the allowable and also the stress in the reinforcement being slightly larger than that calculated by BS 5628; compare line 2 of Fig. 10.14 with Fig. 10.10(c). To determine the strain in the lower reinforcement, using the Eurocode, it would be necessary to know the value of the maximum compressive strain ($\leqslant 0.0035$) and then use the geometry of the figure to calculate the strain at the level of the reinforcement. The calculation can be expressed in the form:

$$\varepsilon_2 = 0.002 - (\varepsilon - 0.002)(7d/3t - 1) \quad (10.17)$$

where ε_2 = strain in the reinforcement at depth d and ε = strain in the upper face of the masonry.

(b) Stress–strain diagram for the reinforcing steel

In the Eurocode the stress–strain relationship for steel is taken as bilinear as shown in Fig. 10.15 rather than the trilinear relationship used in BS 5628 (see Fig. 10.3.).

(c) Conclusion

The main difference between the two codes occurs when the strain distribution is such that the section is in compression throughout. (This

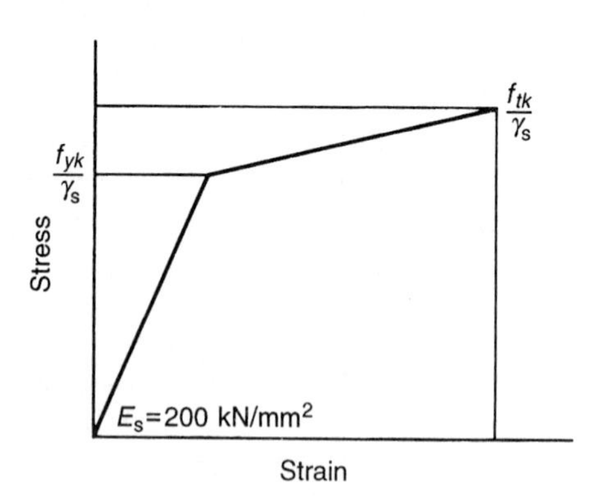

Fig. 10.15 Stress–strain diagram for reinforcement (ENV 1966-1-1).

is illustrated in Fig. 10.10(d) for BS 5628 and Fig. 10.14 (line 2) for ENV 1996-1-1.) Additionally, the method of obtaining the stress, for these cases, will differ because of the different representations of the stress–strain relationship.

For other distributions the design approach for BS 5628 would satisfy the requirements of ENV 1996-1-1 and it is suggested that the methods described in section 10.5 could be used for all cases. No guidance is given in the Eurocode with regard to biaxial bending or slender columns and for these cases the methods described in sections 10.5.3 (b) and 10.5.4 could be used.

11

Prestressed masonry

11.1 INTRODUCTION

Masonry is very strong in compression, but relatively very weak in tension. This restricts its use in elements which are subjected to significant tensile stress. This limitation can be overcome by reinforcing or prestressing. Prestressing of masonry is achieved by applying precompression to counteract, to a desired degree, the tension that would develop under service loading. As a result, prestressing offers several advantages over reinforced masonry, such as the following.

1. *Effective utilization of materials.* In a reinforced masonry element, only the area above the neutral axis in compression will be effective in resisting the applied moment, whereas in a prestressed masonry element the whole section will be effective (Fig. 11.1). Further, in reinforced masonry, the steel strain has to be kept low to keep the cracks within an acceptable limit; hence high tensile steel cannot be used to its optimum.
2. *Increased shear strength.* Figure 11.2 shows the shear strength of reinforced and prestressed brickwork beams with respect to shear arm/effective depth. It is clear that the shear strength of a fully prestressed brickwork beam with bonded tendons is much higher than one of reinforced

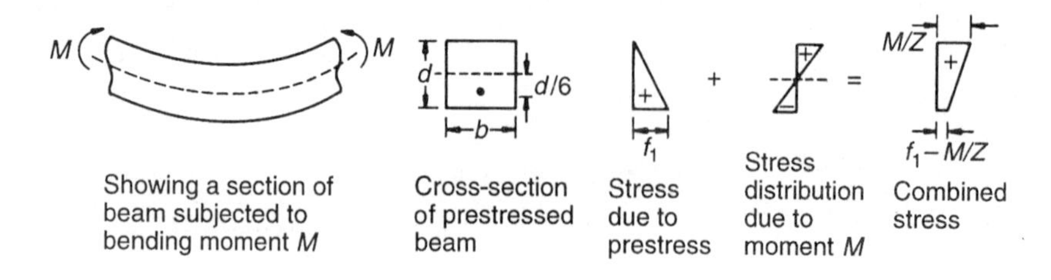

Fig. 11.1 In a prestressed element the whole cross-sectional area is effective in resisting an applied moment.

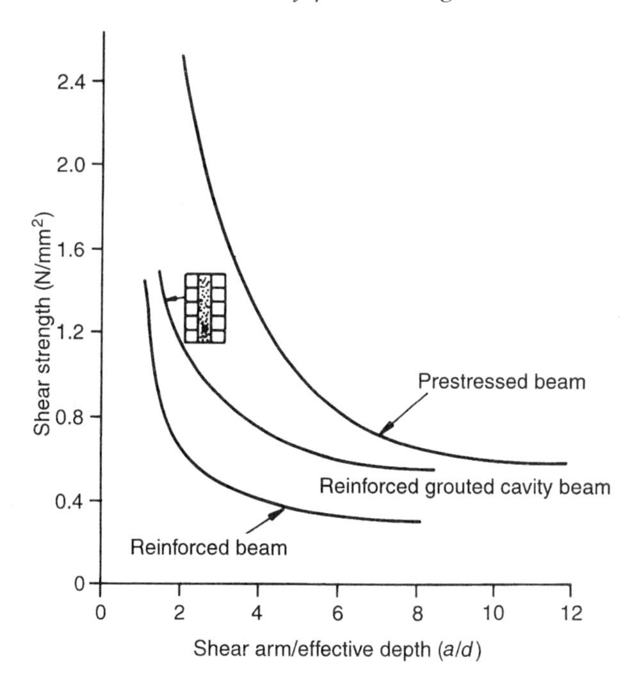

Fig. 11.2 Shear strengths of different types of brickwork beams of similar cross-sections.

brickwork or reinforced grouted brickwork cavity construction. Although the experimental results are for brickwork beams, the findings are applicable also for other type of masonry flexural elements.

3. *Improved service and overload behaviour.* By choosing an appropriate degree of prestressing, cracking and deflection can be controlled. It may, however, be possible to eliminate both cracking and deflection entirely, under service loading in the case of a fully prestressed section. In addition, the cracks which may develop due to overload will close on its removal.
4. *High fatigue resistance.* In prestressed masonry, the amplitude of the change in steel strain is very low under alternating loads; hence it has high fatigue strength.

11.2 METHODS OF PRESTRESSING

The techniques and the methods of prestressing of masonry are similar to those for concrete.

11.2.1 Pretensioning

In this method, the tendons are tensioned to a desired limit between external anchorages and released slowly when both the masonry and its concrete infill have attained sufficient strength. During this operation, the forces in the tendons are transferred to the infill then to the masonry by the bond.

11.2.2 Post-tensioning

In this method, the tendons are tensioned by jacking against the masonry element after it has attained adequate strength. The tendon forces are then transmitted into the masonry through anchorages provided by external bearing plates or set in concrete anchorage blocks. The stresses in anchorage blocks are very high; hence any standard textbook on prestressed concrete should be consulted for their design. In some systems the tendon force is transmitted to the brickwork by means of threaded nuts bearing against steel washers on to a solid steel distributing plate.

The tendons can be left unbonded or bonded. From the point of view of durability, it is highly desirable to protect the tendon by grouting or by other means as mentioned in clause 32.2.6 of BS 5628: Part 2. For brick masonry, post-tensioning will be easier and most likely to be used in practice. It is advantageous to vary the eccentricity of the prestressing force along the length of a flexural member. For example, in a simply supported beam the eccentricity will be largest at the centre where the bending moment is maximum and zero at the support. Unless special clay units are made to suit the cable profile to cater for the applied bending moment at various sections, the use of clay bricks may be limited to:

- Low-level prestressing to increase the shear resistance or to counter the tensile stress developed in a wall due to lateral loading.
- Members with a high level of prestress which carry load primarily due to bending such as beams or retaining walls of small to medium span.

Example 1

A cavity wall brickwork cladding panel of a steel-framed laboratory building (Fig. 11.3) is subjected to the characteristic wind loading of $1.0\,kN/m^2$. Calculate the area of steel and the prestressing force required to stabilize the wall.

Solution

In the serviceability limit state the loads are as follows:

$$\text{design wind load} = \gamma_f w_k = 1 \times 1.0\,kN/m^2$$

($\gamma_f = 1$, clause 20.3.1 BS 5628: Part 2)

$$\text{dead load of the wall} = \gamma_f G_k = 2.6\,\text{kN/m}^2$$

($\gamma_f = 1$, clause 20.3.1 BS 5628: Part 2 and see section 12.2.1)

$$\text{design dead load/metre length of wall} = 2.6 \times 3.6 = 9.1 \times 10^3\,\text{N}$$

$$\text{compressive stress at the base of wall} = 9.1 \times 10^3/\text{area of the wall}$$
$$= 9.1 \times 10^3/1000 \times 102.5$$
$$= 0.089\,\text{N/mm}^2$$

The wall will be treated as a cantilever, which is a safe assumption. Thus

bending moment (BM) at the base of the wall

$$= \frac{1.0 \times 3.6^2}{2} + \frac{1.8 \times 1.0 \times 3.6}{2}$$

$$= 6.5 + 3.3 = 9.8\,\text{kNm/m}$$

Since both walls have the same stiffness,

$$\text{BM/wall} = 9.8/2 = 4.9\,\text{kNm/m}$$

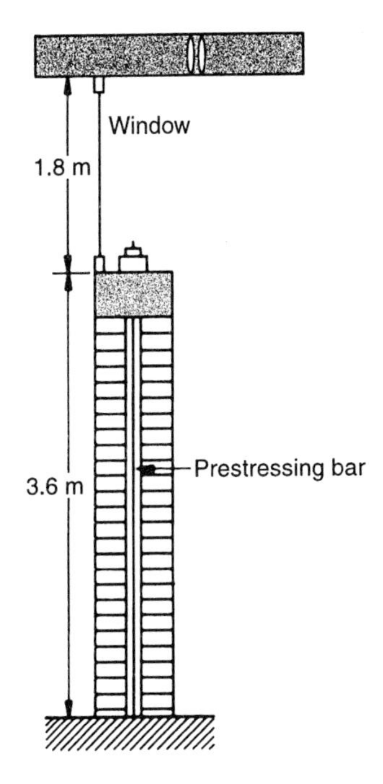

Fig. 11.3 Panel for example 1.

$$\text{stress due to wind loading} = \pm\frac{M}{Z} = \pm\frac{4.9 \times 10^6 \times 6}{1000 \times 102.5^2} = \pm 2.8\,\text{N/mm}^2$$

$$\text{combined stress} = 0.089 - 2.8 = (-)2.71\,\text{N/mm}^2 \quad \text{(tension)}$$

The tension has to be neutralized by the effective prestressing force. Assuming 20% loss of prestress

$$P_e/A = 0.8 \times P_0/A = 2.71$$

Therefore

$$P_0 = (2.71 \times 1000 \times 102.5)/0.8 = 347.2\,\text{kN}$$

$$\text{area of steel required} = (347.2 \times 10^3)/(0.7 \times f_y)$$

$$= (347.2 \times 10^3)/(0.7 \times 1030) = 481.5\,\text{mm}^2$$

Provide one bar of 25 mm diameter ($A_s = 490.6\,\text{mm}^2$).

Alternative solution: If the space is not premium, a diaphragm or cellular wall can be used. The cross-section of the wall is shown in Fig. 11.4. The second moment of area is

$$I_{XX} = 2 \times 615 \times \frac{(102.5)^3}{12} + 2 \times 615 \times 102.5 \times (163.75)^2 + \frac{225^3 \times 100}{12}$$

$$= 110.4 \times 10^6 + 3380 \times 10^6 + 94.9 \times 10^6$$

$$= 3695.7 \times 10^6\,\text{mm}^4$$

$$I_{XX}/\text{m} = \frac{3695.7 \times 10^6}{615} \times 1000 = 6000 \times 10^6\,\text{mm}^4$$

$$\text{area} = 615 \times 2 \times 102.5 + 100 \times 225$$

$$= 126 \times 10^3 + 22.5 \times 10^3$$

$$= 148.5 \times 10^3$$

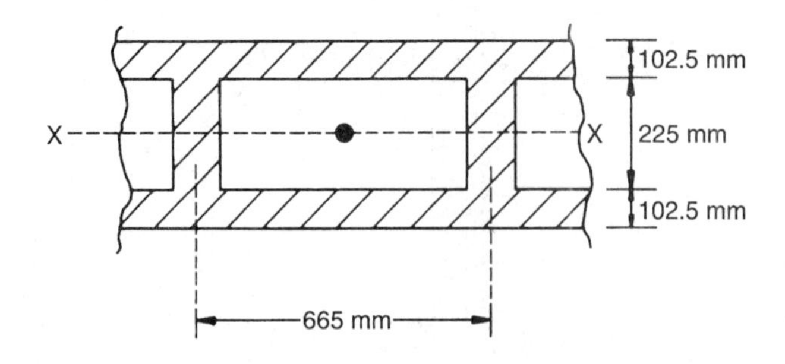

Fig. 11.4 Section of the diaphragm wall for example 1.

$$\text{area/m} = 241.46 \times 10^3$$

$$\text{dead weight} = 2 \times 2.42 \times 3.6 + 2.42 \times 0.225 \times 3.6$$
$$= 17.42 + 1.96$$
$$= 19.38\,\text{kN}$$

$$\text{compressive stress at the base of the wall} = \frac{19.38 \times 10^3}{241.46 \times 10^3} = 0.08\,\text{N/mm}^2$$

The wall will be treated as a cantilever (safe assumption). Then BM at the base of the wall is 9.8 kN m/m and

$$\text{stress due to wind loading} = \pm\frac{9.8 \times 10^6 \times 215}{6000 \times 10^6} = \pm 0.35\,\text{N/mm}^2$$

$$\text{combined stress} = 0.08 - 0.35 = -0.27\,\text{N/mm}^2$$

(about 10 times less than in previous case)

$$P_0 = \frac{0.27 \times 1000 \times 241.46 \times 10^3}{0.8 \times 10^3} = 81.8\,\text{kN}$$

$$\text{area of steel required} = \frac{81.8 \times 10^3}{0.7 \times f_y} = \frac{81.8 \times 10^3}{0.7 \times 1030} = 113.45\,\text{mm}^2$$

Provide one bar of 12 mm diameter.

11.3 BASIC THEORY

The design and analysis of prestressed flexural members is based on the elastic theory of simple bending. The criteria used in the design of such members are the permissible stresses at transfer and at service loads. However, a subsequent check is made to ensure that the member has an adequate margin of safety against the attainment of the ultimate limit state.

11.3.1 Stresses in service

Consider a simply supported prestressed brickwork beam shown in Fig. 11.5(a). The prestressing force P has been applied at an eccentricity of e. Owing to the application of prestress at a distance e, the section is subjected to an axial stress and a hogging moment; the stress distribution is shown in Fig. 11.5(b). As the prestress is applied, the beam will lift upwards and will be subjected to a sagging moment M_i due to its self-weight together with any dead weight acting on the beam at that

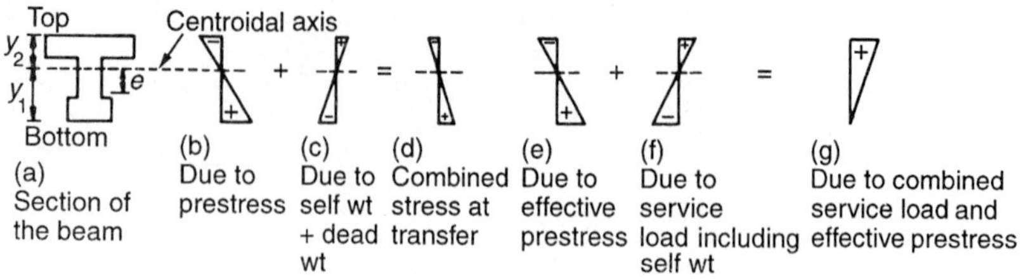

Fig. 11.5 Simply supported prestressed brickwork beam.

time. The stress due to the moment M_{i} is represented in Fig. 11.5(c) and the combined stress due to prestress and moment M_{i} is given in Fig. 11.5(d). At transfer, the tensile stress at the top and compressive stress at the bottom of the section should be less than or equal to the permissible stresses for the brickwork at the critical section. This can be represented in mathematical terms as in the following subsection.

11.3.2 Transfer (initial)

Stress at top

$$f_2 = \frac{P}{A} - \frac{Pey_2}{I} + \frac{M_{\mathrm{i}}}{z_2} \geqslant f_{\mathrm{tt}}$$

$$= \frac{P}{A} - \frac{Pe}{z_2} + \frac{M_{\mathrm{i}}}{z_2} \geqslant f_{\mathrm{tt}} \tag{11.1}$$

Stress at bottom

$$f_1 = \frac{P}{A} + \frac{Pey_1}{I} - \frac{M_{\mathrm{i}}}{z_1} \leqslant f_{\mathrm{ct}}$$

$$= \frac{P}{A} + \frac{Pe}{z_1} - \frac{M_{\mathrm{i}}}{z_1} \leqslant f_{\mathrm{ct}} \tag{11.2}$$

The effective stress distribution due to prestress is shown in Fig. 11.5(e); the stresses will reduce from those shown in (b), because of loss of prestress (to be discussed later in section 11.7). The stresses due to subsequent loading are shown in Fig. 11.5(g) and the resultant stress distribution in Fig. 11.5(h). The governing condition for the design will be that the compressive stress at the top and the tensile stress at the bottom should be less than or equal to the permissible compressive and tensile stresses of the masonry at the critical section (h).

At service

$$\text{stress at top} \quad f_2 = \alpha\left(\frac{P}{A} - \frac{Pey_2}{I}\right) + \frac{M_s}{z_2} \leqslant f_{cs}$$

$$= \alpha(P/A - Pe/z_2) + M_s/z_2 \leqslant f_{cs} \qquad (11.3)$$

and

$$\text{stress at bottom} \quad f_1 = \alpha\left(\frac{P}{A} + \frac{Pey_2}{I}\right) - \frac{M_s}{z_2} \geqslant f_{ts}$$

$$= \alpha(P/A + Pe/z_1) - M_s/z_1 \geqslant f_{ts} \qquad (11.4)$$

From equations (11.1) and (11.3) we get

$$z_2 \geqslant \frac{M_s - \alpha M_i}{f_{cs} - \alpha f_{tt}} \qquad (11.5)$$

$$z_1 \geqslant \frac{M_s - \alpha M_i}{\alpha f_{ct} - f_{ts}} \qquad (11.6)$$

11.3.3 Critical sections

The conditions of equations (11.5) and (11.6) must be satisfied at the critical sections. In a post-tensioned, simply supported masonry beam with curved tendon profile, the maximum bending moment will occur at mid-span, at both transfer and service.

Assuming the values of bending moments M_s, M_{d+L} and M_i all are for mid-span, let

$$M_s = M_{d+L} + M_i \qquad (11.7)$$

Substituting the value of M_s, equations (11.5) and (11.6) become *at transfer*

$$z_2 \geqslant \frac{M_{d+L} + (1-\alpha)M_i}{f_{cs} - \alpha f_{tt}} \qquad (11.8)$$

$$z_1 \geqslant \frac{M_{d+L} + (1-\alpha)M_i}{\alpha f_{ct} - f_{ts}} \qquad (11.9)$$

In prestressed or post-tensioned fully bonded beams with straight tendons the critical sections of the beam at transfer will be near the ends. At the end of the beam, moment M_i may be assumed to be zero.

Substituting the value of M_i in equations (11.5) and (11.6)

$$z_2 \geqslant \frac{M_{d+L} + M_i}{f_{cs} - \alpha f_{tt}} \tag{11.10}$$

$$z_1 \geqslant \frac{M_{d+L} + M_i}{\alpha f_{ct} - f_{ts}} \tag{11.11}$$

Depending on the chosen cable profiles, the values of z_1 and z_2 can be found from the equations (11.8) to (11.11).

Having found the values of z_1 and z_2 the values of prestressing force and the eccentricity can be found from equations (11.1) and (11.4) as

$$P_{min} = \frac{[(M_s - \alpha M_i) + (z_1 f_{ts} + \alpha z_2 f_{tt})]A}{\alpha(z_1 + z_2)} \tag{11.12}$$

$$e_{max} = \frac{z_2}{A} + \frac{M_i - z_2 f_{tt}}{P} \tag{11.13}$$

11.3.4 Permissible tendon zone

The prestressing force will be constant throughout the length of the beam, but the bending moment is variable. As the eccentricity was calculated from the critical section, where the bending moment was maximum, it is essential to reduce it at various sections of the beam to keep the tensile stresses within the permissible limit. Since the tensile stresses become the critical criteria, using equations (11.1) and (11.4), we get

$$e_1(\text{lower limit}) \leqslant \frac{z_2}{A} + \frac{M_i - z_2 f_{tt}}{P} \tag{11.14}$$

$$e_2(\text{upper limit}) \leqslant -\frac{z_1}{A} - \frac{M_s - z_1 f_{ts}}{\alpha P} \tag{11.15}$$

At present, in a prestressed masonry beam, no tension is allowed and since the bending moment due to self-weight will be zero at the end, the lower limit of eccentricity from equation (11.14) will become

$$e_1 \leqslant z_2/A \tag{11.16}$$

where z_2/A is the 'kern' limit.

In the case of a straight tendon this eccentricity will govern the value of prestressing force, and hence from equations (11.1) and (11.4), P can be obtained as

$$P = \frac{M_s A}{\alpha(z_1 + z_2)} \tag{11.17}$$

Example 2

A post-tensioned masonry beam (Fig. 11.6) of span 6 m, simply supported, carries a characteristic superimposed dead load of 2 kN/m and a characteristic live load of 3.5 kN/m. The masonry characteristic strength $f_k = 19.2\ \text{N/mm}^2$ at transfer and service, and the unit weight of masonry is $21\ \text{kN/m}^3$. Design the beam for serviceability condition ($\gamma_f = 1$).

Solution

$$f_{tc} = 0.5 \times 19.2 = 9.6\ \text{N/mm}^2$$

(clause 29.1, BS 5628: Part 2)

$$f_{cs} = 0.4 \times 19.2 = 7.68\ \text{N/mm}^2$$

(clause 29.2, BS 5628: Part 2)

$$f_{tt} = f_{ts} = 0$$

$$M_{d+L} \frac{(2 + 3.5) \times 6^2}{8} = 24.75\ \text{kN}$$

Assume M_i is 30% of M_{d+L} so

$$M_i = 0.3 \times 24.75 = 7.425\ \text{kN m}$$

$$z_2 \geqslant \frac{(24.75 + 7.425) \times 10^6}{7.68 - 0} = 4.19 \times 10^6$$

(from equation (11.10))

$$z_1 \geqslant \frac{(24.75 + 7.425) \times 10^6}{0.8 \times 9.6} = 4.18 \times 10^6$$

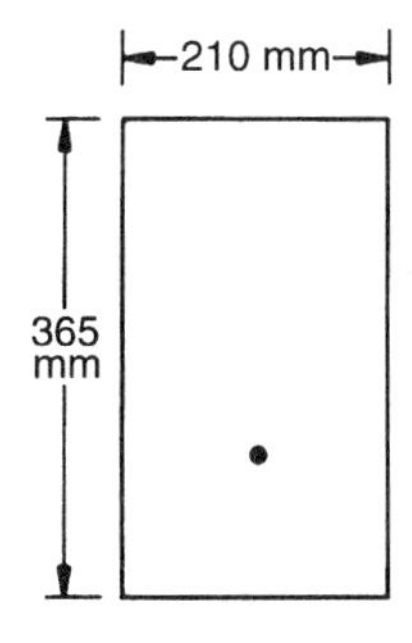

Fig. 11.6 Cross-section of the beam for example 2.

$$\text{loss ratio} = \alpha = \frac{\text{effective prestress}}{\text{prestress at transfer}} = 0.8 \quad \text{(assumed)}$$

Assume rectangular section

$$d = \left(\frac{4.19 \times 10^6 \times 6}{b}\right)^{1/2} = \left(\frac{4.19 \times 10^6 \times 6}{210}\right)^{1/2} = 346\,\text{mm}$$

Provide $d = 365$ mm to take into account the thickness of a brick course. Correct value of M_i is

$$M_i = \frac{0.210 \times 0.365 \times 21 \times 6^2}{8} = 7.24\ \text{kN m} < 7.425 \quad \text{(assumed)}$$

For straight tendon,

$$e = z_2/A = bd^2/(6bd) = d/6 \quad \text{(from equation (11.16))}$$

$$= 365/6 = 60.83 \quad \text{(from equation (11.17))}$$

$$P = \frac{M_s A}{\alpha(z_1 + z_2)}$$

$$= \frac{(24.75 + 7.24) \times 10^6 \times 6}{0.8 \times 2 \times d}$$

$$= \frac{31.99 \times 10^6 \times 6}{0.8 \times 2 \times 365 \times 10^3} = 328.7\ \text{kN}$$

11.4 A GENERAL FLEXURAL THEORY

The behaviour of prestressed masonry beams at ultimate load is very similar to that of reinforced masonry beams discussed in Chapter 10. Hence, a similar approach as applied to reinforced masonry with a slight modification to find the ultimate flexural strength of a prestressed masonry beam is used. For all practical purposes, it is assumed that flexural failure will occur by crushing of the masonry at an ultimate strain of 0.0035, and the stress diagram for the compressive zone will correspond to the actual stress–strain curve of masonry up to failure.

Now, let us consider the prestressed masonry beam shown in Fig. 11.7(a). For equilibrium, the forces of compression and tension must be equal, hence

$$\lambda_1 f_m b d_c = A_{ps} f_{su} \quad \text{or} \quad f_{su} = \frac{\lambda_1 f_m b d_c}{A_{ps}} \tag{11.18}$$

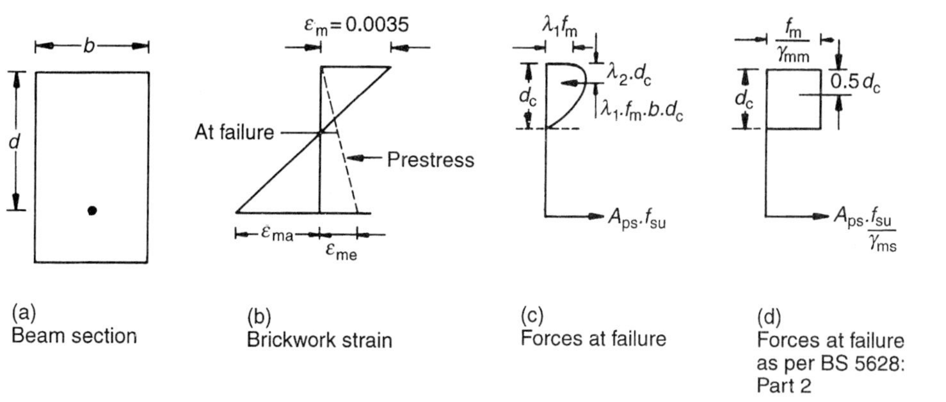

Fig. 11.7 The strain and stress block at failure of a prestressed masonry beam.

The ultimate strain ε_{su} in prestressing steel consists of strain due to prestress and applied load, hence

$$\varepsilon_{su} = \varepsilon_{sa} + \varepsilon_{se} \tag{11.19}$$

where ε_{sa} is due to applied load and ε_{se} is due to effective prestress after losses. From Fig. 11.7(b) it can be seen that strain due to the applied load is equal to

$$\varepsilon_{sa} = \varepsilon_{ma} + \varepsilon_{me}$$

where

$$\varepsilon_{me} = \frac{\text{prestressing stress at tendon level}}{E_m} \tag{11.20}$$

Assuming full bond exists between the steel, grout and masonry at failure, then the strain in steel may be given by

$$\varepsilon_{sa} = \varepsilon_m\left(\frac{d - d_c}{d_c}\right) + \varepsilon_{me} \tag{11.21}$$

Substituting the value of ε_{sa} from equation (11.21) into equation (11.19), we get

$$\varepsilon_{su} = \varepsilon_m\left(\frac{d - d_c}{d_c}\right) + \varepsilon_{me} + \varepsilon_{se}$$

or

$$d_c = \frac{\varepsilon_m}{\varepsilon_{su} + \varepsilon_m - \varepsilon_{me} - \varepsilon_{se}}\, d \tag{11.22}$$

Combining equations (11.18) and (11.22) gives

$$f_{su} = \frac{\lambda_1 f_m bd}{A_{ps}} \frac{\varepsilon_m}{\varepsilon_{su} + \varepsilon_m - \varepsilon_{me} - \varepsilon_{se}}$$

$$= \frac{\lambda_1 f_m}{\rho} \frac{\varepsilon_m}{\varepsilon_{su} + \varepsilon_m - \varepsilon_{me} - \varepsilon_{se}} \qquad (11.23)$$

At the ultimate limit state, the values of f_{su} and ε_{su} must satisfy equation (11.23) and also define a point on the stress–strain curve for the steel (see Fig. 2.7). Having found f_{su} and the tendon strain ε_{su}, the depth of the neutral axis d_c can be obtained from equation (11.22). The ultimate moment of resistance is then

$$M_u = A_{su} f_{su}(d - \lambda_2 d_c)$$

$$= A_{su} f_{su}\left(1 - \rho \frac{\lambda_2 f_{su}}{\lambda_1 f_m}\right)d \qquad (11.24)$$

Generally, an idealized stress block is used for design purposes. Figure 11.7(d) shows the rectangular stress block suggested in the British Code of Practice for prestressed masonry. The values of λ_1 and λ_2 corresponding to this stress block are 1 and 0.5.

The materials partial safety factors are γ_{mm} for masonry and γ_{ms} for steel. The general flexural theory given in this section can easily be modified to take account of these.

Example 3

A bonded post-tensioned masonry beam of rectangular cross-section 210 × 365 mm as shown in Fig. 11.8 has been prestressed to effective stress of 900 N/mm^2 by four 10.9 mm diameter stabilized strands of characteristic strength of 1700 N/mm^2. The area of steel provided is 288mm^2. The initial modulus of elasticity of the steel is 195 kN/mm^2 and the stress–strain relationship is given in Fig. 2.7. The masonry in 1:$\frac{1}{4}$:3 mortar has a characteristic strength parallel to the bed joint of 21 N/mm^2 and modulus of elasticity 15.3 kN/mm^2.

Using the simplified stress block of BS 5628: Part 2, calculate the ultimate moment of resistance of the beam.

Solution

We have

$$\text{area } A = 210 \times 365 = 76\,650\,\text{mm}^2$$

$$I = bd^3/12 = 210 \times (365)^3/12 = 8.5 \times 10^8\,\text{mm}^4$$

$$P_e = 900 \times 288 = 259.2 \times 10^3\,\text{N}$$

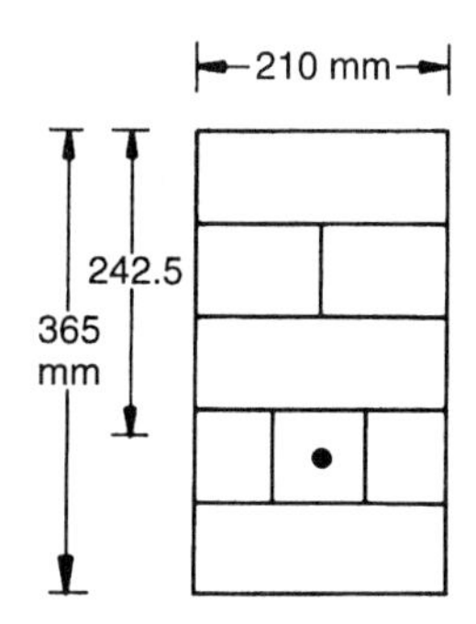

Fig. 11.8 Cross-section of the beam for example 3.

Masonry strain at tendon level

$$\varepsilon_{me} = \frac{1}{E_m}\left(\frac{P}{A} + \frac{Pey}{I}\right)$$

$$= \frac{259.2 \times 10^3}{15.3 \times 10^3}\left(\frac{1}{76\,650} + \frac{60 \times 60}{8.5 \times 10^8}\right)$$

$$= 0.000\,293$$

From equation (11.22),

$$\varepsilon_{su} = \varepsilon_m\left(\frac{d - d_c}{d_c}\right) + \varepsilon_{me} + \varepsilon_{se}$$

$$= 0.0035[(242.5 - d_c)/d_c)] + 0.000\,293$$

$$+ 900/(195 \times 10^3)$$

$$= 0.8488/d_c + 0.0035 + 0.000\,293 + 0.004\,615$$

$$= 0.8488/d_c + 0.001\,41$$

Therefore

$$d_c = \frac{0.8488}{\varepsilon_{su} - 0.001\,41}$$

From equation (11.18) and using the stress block of BS 5628: Part 2

$$f_m b d_c / \gamma_{mm} = A_{ps} f_{su}$$

$$(1 \times 21/2)\ 210 d_c = 288 f_{su}$$

or

$$f_{su} = \frac{21 \times 210}{2 \times 288} d_c = 7.66 d_c$$

Therefore

$$f_{su} = \frac{7.66 \times 0.8488}{\varepsilon_{su} - 0.001\,41} = \frac{6.5}{\varepsilon_{su} - 0.001\,41}$$

This is solved with the stress–strain curve given in Fig. 2.7.

$$f_{su} = 1214\ \text{N/mm}^2 \quad \varepsilon_{su} = 0.006\,76$$

Therefore

$$d_c = 1214/7.66 = 158.5\ \text{mm}$$

From equation (11.24)

$$M_u = A_{su} f_{su} (d - \lambda_2 d_c) \quad \text{where } \lambda_2 = 0.5\ \text{(BS 5628)}$$

$$= 288 \times 1214\,(242.5 - 0.5 \times 158.5)\,/10^6\ \text{kN m}$$

$$= 57.0\ \text{kN m}$$

11.5 SHEAR STRESS

The shear stress due to the loading must be checked to ensure that the value is within the acceptable limit. The characteristic shear strength with bonded tendons for elements prestressed parallel to the bed joint should be taken as $0.35\ \text{N/mm}^2$. The characteristic shear strength for prestressed elements with bonded tendons, where prestressing is normal to the bed joint, can be obtained from

$$f_v = 0.87 + 0.21\, g_b$$

where g_b is the prestressing stress. The maximum value should not exceed $1.75\ \text{N/mm}^2$.

The prestressed elements with unbonded tendons have much lower strength than with bonded tendons. The value given by the equation above is quite different from the recommendation of BS 5628: Part 2, which does not differentiate between bonded and unbonded tendons. This may not be correct according to the limited experimental results at present available.

11.6 DEFLECTIONS

In the design of a prestressed member, both short- and long-term deflections need to be checked. The short-term deflection is due to the prestress, applied dead and live loads. The effect of creep increases the deflection in the long term, and hence this must be taken into consideration. The long-term deflection will result from creep under prestress and dead weight, i.e. permanent loads acting on the member plus the live load. If part of the live load is of a permanent nature, the effect of creep must be considered in the design. The deflections under service loading should not exceed the values given in the code of practice for a particular type of beam. The code, at present, does not allow any tension; hence the beam must remain uncracked. This makes deflection calculation much easier. However, the deflection of a prestressed beam after cracking and up to failure can be easily calculated by the rigorous method given elsewhere (Pedreschi and Sinha, 1985).

Example 4

The beam of example 3 is to be used as simply supported on a 6 m span. It carries a characteristic superimposed dead load of 2 kN/m^2 and live load of 3.0 kN/m^2; 50% of the live load is of permanent nature. Calculate the short- and long-term deflection.

Solution

We have

$$\text{dead weight moment of the beam} = \frac{0.21 \times 0.365 \times 21 \times 6^2}{8} = 7.24\ \text{kN m}$$

$$\text{BM due to live + dead weight} = \frac{(2+3) \times 6^2}{8} = 22.5\ \text{kN}$$

$$\text{total applied moment} = 29.74\ \text{kN m}$$

$$\text{tensile stress} = 10^6 \times \frac{29.74 \times 6}{bd^2} = 10^6 \times \frac{29.74 \times 6}{210 \times (365)^2} = 6.38\ \text{N/mm}^2$$

$$\text{compressive stress due to effective prestress} = \frac{P_e}{A} + \frac{P_e e y}{I}$$

$$= \frac{259.2 \times 10^3}{210 \times 365} + \frac{259.2 \times 10^3 \times 182.5 \times 60}{210 \times (365)^3/12}$$

$$= 3.38 + 3.34 = 6.72\ \text{N/mm}^2 > 6.38\ \text{N/mm}^2$$

Hence the beam will remain uncracked.

The short-term deflection is calculated as follows.

$$\text{deflection due to prestress} = (-)\frac{ML^2}{8EI} = (-)\frac{P_e e L^2}{8EI} \quad \text{(hogging)}$$

$$= (-)\frac{259.2 \times 10^3 \times 60 \times 36 \times 10^6 \times 12}{8 \times 15.3 \times 10^3 \times 210 \times (365)^3} = (-)\,5.38\ \text{mm}$$

Deflection due to self-weight + dead weight + 50% live load, taking $\gamma_f = 1$, is

$$\text{deflection due to loads} = +\frac{5\,wL^4}{384\,EI} \quad \text{(sagging)}$$

$$= +\frac{5 \times (1.61 + 1.5 + 2.0) \times (6000)^4 \times 12}{384 \times 15.3 \times 10^3 \times 210 \times (365)^3}$$

$$= 6.62 \quad \text{(sagging)}$$

deflection due to live load

$$= +\frac{5 \times 1.5 \times (6000)^4 \times 12}{384 \times 15.3 \times 10^3 \times 210 \times (365)^3} = 1.94 \quad \text{(sagging)}$$

Hence

$$\text{short-term deflection} = -5.38 + 6.62 + 1.94 = 3.18\ \text{mm}$$

The long-term deflection is given by

$$\text{long-term deflection} = (\text{short-term deflection due to prestress} + \text{dead weight})\,(1 + \phi) + \text{live load deflection}$$

where ϕ is the creep factor from BS 5628: Part 2, $\phi = 1.5$. Hence

$$\text{long-term deflection} = (-5.38 + 6.62)\,(1 + 1.5) + 1.94 = 5.04\ \text{mm.}$$

11.7 LOSS OF PRESTRESS

The prestress which is applied initially is reduced due to immediate and long-term losses. The immediate loss takes place at transfer due to elastic shortening of the masonry, friction and slip of tendons during the anchorage. The long-term loss occurs over a period of time and may result from relaxation of tendons, creep, shrinkage and moisture movement of brickwork.

11.7.1 Elastic shortening

When the forces from the external anchorages are released on to the member to be prestressed, they cause elastic deformation, i.e. shortening of the masonry or surrounding concrete as the case may be. This will cause reduction of stress in the tendon as the strain in the surrounding concrete or brickwork must be equal to the reduction of strain in the tendon.

In a pretensioned member, the force P_0 required in the tendon prior to elastic shortening can be calculated as explained below. Let us assume that P_0 = force immediately before transfer, P_i = force in tendon after elastic shortening, E_m and E_s = Young's modulus of elasticity for masonry and steel, $\Delta\sigma_s$ = decrease of stress in tendon, f_m' = masonry compressive stress at tendon level after transfer, A = cross-sectional area of beam and A_{ps} = area of prestressing steel. Hence,

$$\frac{\Delta\sigma_s}{E_s} = \frac{f_m'}{E_m} \tag{11.25}$$

or

$$\Delta\sigma_s = \frac{E_s}{E_m} f_m' \qquad \frac{E_s}{E_m} = m \quad \text{(modular ratio)}$$

$$\Delta\sigma_s = m f_m'$$

From equilibrium

$$f_m' = \left(\frac{P_i}{A} + \frac{P_i e^2}{I}\right) = P_i\left(\frac{1}{A} + \frac{e^2}{I}\right) \tag{11.26}$$

where e is tendon eccentricity. Also,

$$\Delta\sigma_s = \frac{P_0 - P_i}{A_{ps}} \tag{11.27}$$

From (11.25), (11.26) and (11.27)

$$P_0 = P_i\left[1 + \left(\frac{1}{A} + \frac{e^2}{I}\right) m A_{ps}\right]$$

$$= P_i\left[1 + \left(\frac{1}{A} + \frac{e^2}{I}\right)\frac{E_s}{E_m} A_{ps}\right] \tag{11.28}$$

In post-tensioning, the tendon is stretched against the masonry member itself. Thus the masonry is subjected to elastic deformation during the post-tensioning operation and the tendon is locked off when desired prestress or elongation of tendon has been achieved. Thus in a post-tensioned member with single tendon or multiple tendons, there will be no loss due to elastic shortening provided all of them are stretched simultaneously. If the tendons are stretched in a sequence, there will be loss of prestress in the tendon or tendons which were already stressed.

11.7.2 Loss due to friction

As the prestressing force is determined from the oil pressure in the jack, the actual force in the tendon will be reduced by friction in the jack. Data to allow for this may be obtained from the manufacturer of the particular jacking system in use.

During post-tensioning operations, there will be a further loss of prestress because of friction between the sides of the duct and the cable. The loss in the transmitted force increases as the distance increases from the jacking end and can be represented by:

$$P_x = P_0 e^{-kx}$$

where P_x = force at distance x from the stressing anchorage, k = coefficient depending on the type of duct, x = distance from the jack, P_0 = force at the stressing anchorage and e = base of Napierian logarithms.

In masonry with a preformed cavity to accommodate straight tendons, the loss will be negligible as the tendons seldom touch the sides of the member.

11.7.3 Loss due to slip in anchorages

The anchorage fixtures are subjected to stress at transfer and will deform. As a result, the frictional wedges used to hold the cables slip a little distance which can vary from 0 to 5 mm. This causes a reduction in prestress which may be considerable in a short post-tensioned member. The loss cannot be predicted theoretically but can only be evaluated from the data obtained from the manufacturer of the anchorage system. However, in practice, this loss can be completely eliminated at the dead end by stressing the tendon and releasing the prestressing force without anchoring at the jacking end or can be compensated by overstressing. No loss of prestress occurs in a system which uses threaded bar and nuts for post-tensioning.

11.7.4 Relaxation loss

Relaxation loss can be defined as loss of stress at constant strain over a period of time. This loss in prestress depends upon the initial stress and the type of steel used. Normally, the test data for 1000 hours relaxation loss at an ambient temperature of 20°C will be available, for an initial load of 60%, 70% and 80% of the breaking load, from the manufacturers of the prestressing steel. Linear interpolation of this loss between 60% and 30% of breaking load is allowed, assuming that the loss reduces to zero at 30% of the breaking load. The value of the initial force is taken immediately after stressing in the case of pretensioning and at transfer for post-tensioning. In absence of data, the values given in appropriate codes should be used for the design.

The relaxation, shrinkage and creep losses are interdependent, and hence in prestressed concrete the 1000 hour test value is multiplied by a relaxation factor to take these together into account. However, no such data are available for brickwork; hence the total loss will be overestimated, if each is added separately.

11.7.5 Loss due to moisture expansion, shrinkage and creep

The effect of moisture expansion of fired clay bricks will be to increase the prestressing force in tendons, but this is disregarded in design. However, if the moisture movement causes shrinkage in masonry, there will be a loss of prestress. The code recommends a value of maximum strain of 500×10^{-6} for calcium silicate and concrete bricks. The loss of prestress can be calculated from the known value of strain.

Rather limited data are available for determination of loss of prestress due to creep in brickwork. The code recommends the creep strain is equal to 1.5 times the elastic strain for brickwork and 3 times for concrete blockwork and these values should be used for the design in the absence of specific data.

11.7.6 Thermal effect

In practice, materials of different coefficients of thermal expansion are used and this must be considered in the design. In closed buildings, the structural elements are subjected to low temperature fluctuations, but this is not the case for the external walls, especially prestressed wide-cavity cellular walls where the temperatures of the inner and outer walls will always be quite different. An unbounded tendon in a cavity will generally be at a different temperature from the inner or outer wall which may result in loss of prestress. Such effects are, however, difficult to estimate.

12

Design calculations for a seven-storey dormitory building according to BS 5628

12.1 INTRODUCTION

As an illustration of structural design calculations based on BS 5628 we may consider a building having the layout shown in Figs. 12.1 and 12.2. It is assumed that the roof and floor slabs are of continuous *in situ* reinforced concrete construction. The structure has been kept simple to show the principle of limit state design. Only two walls above G.L. – an internal wall A, heavily loaded compared to the others, and a cavity wall B, have been considered. The inner leaf is assumed to support its own weight together with roof and floor loads, whilst the outer leaf will

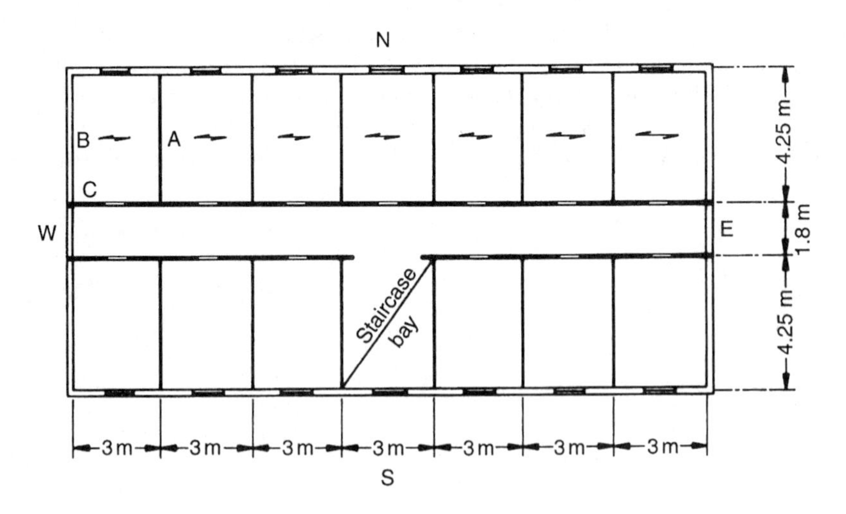

Fig. 12.1 Typical plan of a building.

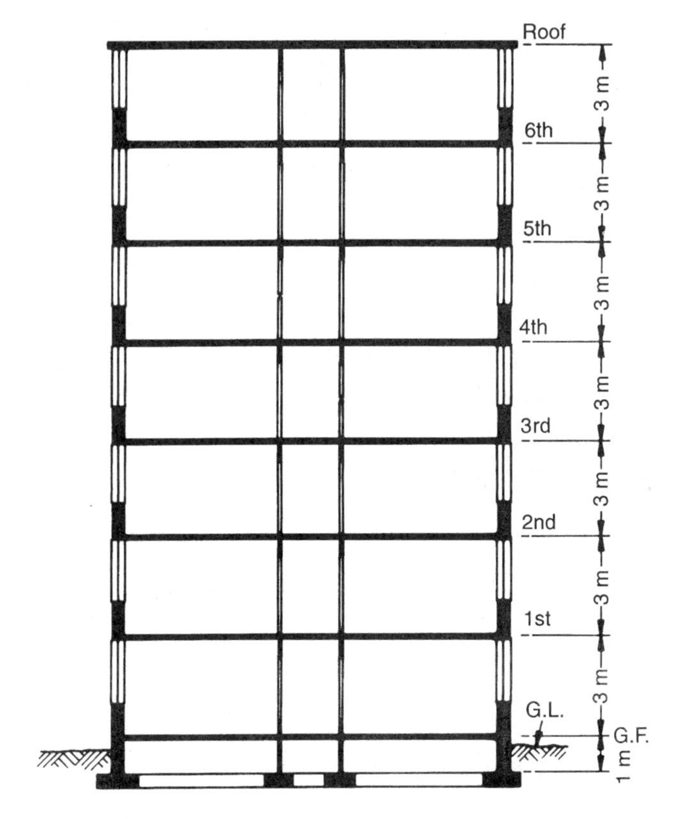

Fig. 12.2 Typical section of a building.

support only its own weight. The design loads and design assumptions are given in section 12.2.

12.2 BASIS OF DESIGN: LOADINGS

- Roof: dead weight, 3.5 kN/m^2
  imposed load, 1.5 kN/m^2
- Floor: dead weight including finishings and partition, 4.8 kN/m^2
  (see section 12.10 for sample calculation)
- Imposed load, 1.5 kN/m^2
- Wall: 102.5 mm with 13 mm plaster both sides, 2.6 kN/m^2
  102.5 mm and inner skin of 255 mm cavity wall, 2.42 kN/m^2
  (i.e. 102.5 mm + one face plaster)
- Wind load: speed (Edinburgh area), 50 m/s

12.3 QUALITY CONTROL : PARTIAL SAFETY FACTORS

Assume normal quality control both at the factory and on site. The partial safety factors for the materials are

$$\gamma_m = 3.5 \quad \text{(table 4, BS 5628)}$$

$$\gamma_{mv} = 2.5 \quad \text{(clause 27.4)}$$

12.4 CALCULATION OF VERTICAL LOADING ON WALLS

12.4.1 Loading on internal wall A

The loading on this wall is summarized in Table 12.1.

12.4.2 Loading on external cavity wall B

(a) Inner leaf

The loading on the inner leaf of this wall is shown in Table 12.2.

(b) Outer leaf

For the outer leaf of this wall

$$\text{load/m at floor} = 2.42 \times 3 = 7.26\,\text{kN/m}$$

$$\text{imposed load} = 0$$

12.4.3 Total dead weight of the building above GL

Neglecting openings, etc., we have

$$\begin{aligned} G_k &= 3.5 \times 21 \times 10.5 + 6 \times 4.8 \times 21 \times 10.5 \\ &\quad + (12 \times 2.6 \times 4.25 \times 2.85 + 4 \times 2 \times 2.42 \times 4.25 \times 2.85 \\ &\quad + 2 \times 21 \times 2.6 \times 2.85 + 21 \times 2 \times 2 \times 2.42 \times 2.85) \times 7 \\ &= 17\,643\,\text{kN} \end{aligned}$$

12.5 WIND LOADING

12.5.1 General stability

To explain the method, only walls A and B are considered in the calculation; hence wind blowing from either north or south direction is critical and evaluated. In the east–west direction the cavity and corridor walls

Table 12.1 Loading on wall A per metre run

Calculation for floor level considered		*Load/m run* (kN/m)		
		Dead load at floor	*Cumulative dead load to floor*, G_k	*Cumulative live load to floor* Q_k
6th floor				
roof dead weight, $3.5 \times 3 \times 1.2$[a]	$= 12.6$			
weight of wall, 2.6×2.85	$= 7.4$			
	20.0 kN/m	20.0	20.0	5.4
imposed load, $1.5 \times 3 \times 1.2$[a]	$= 5.4$ kN/m			
5th floor				
floor dead weight, $4.8 \times 3 \times 1.2$[a]	$= 17.28$			
wall	$= 7.40$			
	24.68 kN/m	24.68	44.68	9.72
90% of imposed load, $2 \times 5.4 \times 0.9$	$= 9.72$ kN/m			
4th floor				
floor dead weight, $4.8 \times 3 \times 1.2$[a]	$= 17.28$			
wall	$= 7.40$			
	24.68 kN/m	24.68	69.36	12.96
80% of 3 floors imposed load $3 \times 5.4 \times 0.8 = 12.96$ kN/m				

Table 12.1 (*Contd*)

Calculation for floor level considered			*Load/m run* (kN/m)		
			Dead load at floor	*Cumulative dead load to floor,* G_k	*Cumulative live load to floor* Q_k
3rd floor					
floor dead weight, $4.8 \times 3 \times 1.2$[a]	=	17.28			
wall	=	7.40			
		24.68 kN/m	24.68	94.04	15.12
70% of 4 floors imposed load, $4 \times 5.4 \times 0.7$	=	15.12 kN/m			
2nd floor					
floor dead weight, $4.8 \times 3 \times 1.2$[a]	=	17.28			
wall	=	7.40			
		24.68 kN/m	24.68	118.72	16.2
60% of 5 floors imposed load, $5 \times 5.4 \times 0.6$	=	16.2 kN/m			

1st floor				
floor dead weight, $4.8 \times 3 \times 1.2$[a] =	17.28			
wall =	7.40			
	24.68 kN/m	24.68	143.4	19.44
60% of 6 floors imposed load, $6 \times 5.4 \times 0.6$ =	19.44 kN/m			
Ground floor				
floor dead weight, $4.8 \times 3 \times 1.2$[a] =	17.28			
wall =	7.40			
	24.68 kN/m	24.68	168.08	22.68
60% of 7 floors imposed load, $7 \times 5.4 \times 0.6$ =	22.68 kN/m			

[a] The factor 1.2 comes from table 3.6, BS 8110: Part 1: 1985.

Table 12.2 Loading on wall B per metre run; inner leaf

Calculation for floor level considered			*Load/m run* (kN/m)		
			Dead load at floor	*Cumulative dead load to floor, G_k*	*Cumulative live load to floor, Q_k*
6th floor					
roof dead weight, $3.5 \times 3 \times 0.45$[a]	=	4.725			
wall (roof to 6th floor), 2.42×2.85	=	6.897			
		11.62 kN/m	11.62	11.62	2.025
5th floor					
floor dead weight, $4.8 \times 3 \times 0.45$[a]	=	6.48			
wall 6th to 5th	=	6.897			
		13.38 kN/m	13.38	25.0	3.645
90% of imposed load, $2 \times 2.025 \times 0.9$	=	3.645 kN/m			
4th floor					
dead weight same as 5th			13.38	38.38	4.86
80% of 3 floors imposed load, $3 \times 2.025 \times 0.8$	=	4.86 kN/m			

3rd floor				
dead weight same as 5th		13.38	51.76	5.67
70% of 4 floors imposed load, 4 × 2.025 × 0.7	= 5.67 kN/m			
2nd floor				
dead weight same as 5th		13.38	65.14	6.08
60% of 5 floors imposed load, 5 × 2.025 × 0.6	= 6.08 kN/m			
1st floor				
dead weight same as 5th		13.38	78.54	7.29
60% of 6 floors imposed load, 6 × 2.025 × 0.6	= 7.29 kN/m			
Ground floor				
dead weight same as 5th		13.38	91.90	8.51
60% of 7 floors imposed load, 7 × 2.025 × 0.6	= 8.51 kN/m			

[a] Factor 0.45 from table 3.6, BS 8110: Part 1:1985.
[b] Imposed load reduction from table 2, BS 6399: Part 1.

will provide the resistance to wind loading. In an actual design, the designer must of course check that the structure is safe for wind blowing east–west and vice versa.

In the calculation below it has further been assumed that the walls act as independent cantilevers; and hence moments or forces are apportioned according to their stiffness.

12.5.2 Wind loads

These are calculated according to CP 3, Chapter V: Part 2. We have

$$V_s = VS_1S_2S_3$$

$$S_1 = S_3 = 1.0$$

Using ground roughness category 3, Class B, with height of the building = 21.0 m, from Table 3, CP 3, Chapter V: Part 2

$$S_2 = 0.91$$

Therefore design wind speed is

$$V_s = 50 \times 1 \times 1 \times 0.91 = 45.5\,\text{m/s}$$

and dynamic wind pressure is

$$q = 0.613 \times (45.5)^2 = 1269.0\,\text{N/m}^2$$

From Clause 7.3, CP 3, Chapter V: Part 2, total wind force

$$F = C_f q A_e \quad (C_f = 1.1, \text{Table 10})$$

$$A_e = \text{effective surface area}$$

The total maximum bending moment is

$$\text{total max. BM} = F \times h/2$$

where h is the height under consideration. Total BM just above floor level is given for each floor by:

- 6th floor
$$C_f q A_e \times h/2 = 1.1 \times (1269/10^3) \times 21 \times 3 \times 3/2 = 131.9\,\text{kN m}$$
- 5th floor
$$1.1 \times (1269/10^3) \times 21 \times 6 \times 3 = 527.6\,\text{kN m}$$
- 4th floor
$$(1.1 \times 1269 \times 21/10^3) \times 9 \times 9/2 = 1187.20\,\text{kN m}$$
- 3rd floor
$$29.313 \times (12 \times 12/2) = 2110.54\,\text{kN m}$$
- 2nd floor
$$29.313 \times (15 \times 15/2) = 3297.70\,\text{kN m}$$

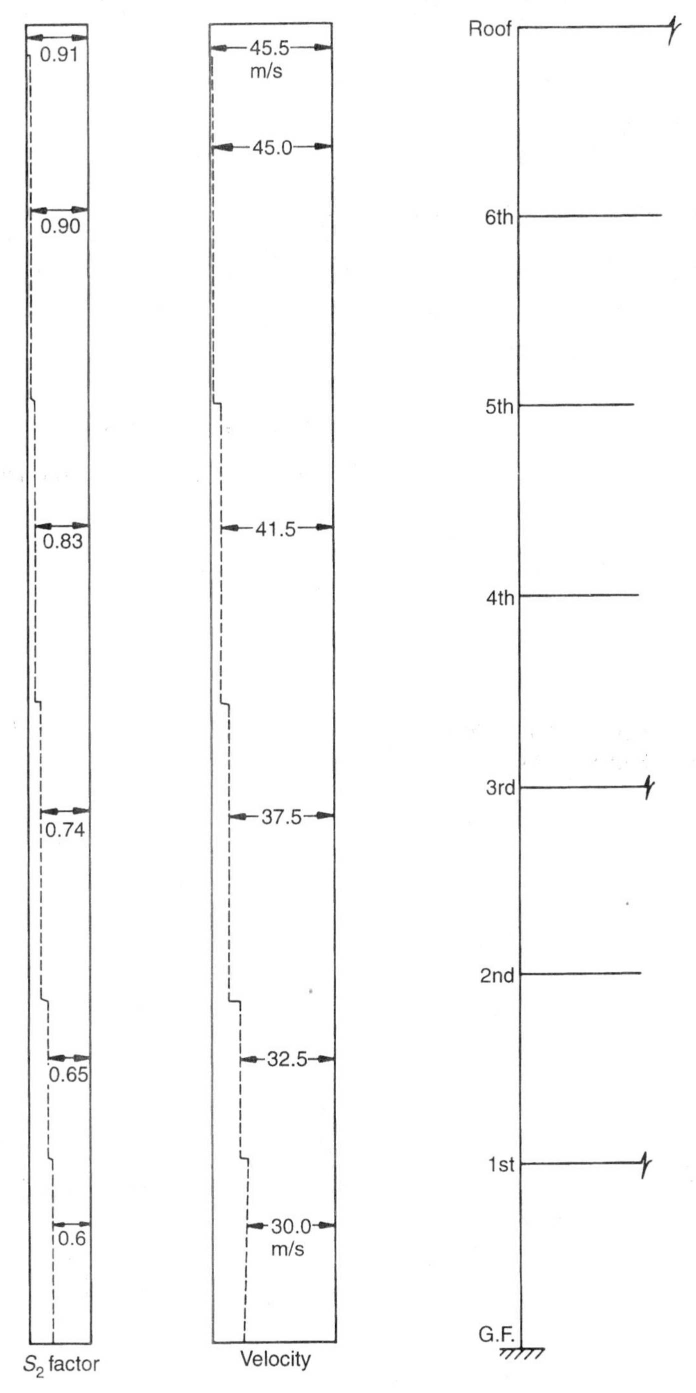

Fig. 12.3 The variation of the factor S_2 and the wind velocity along the height of the building. (Assumptions made in the design shown in full lines.)

- 1st floor

$$29.313 \times (18 \times 18/2) = 4748.71\,\text{kN m}$$

- ground floor

$$1.1 \times (1269/10^3) \times 21 \times 21 \times 21/2 = 6463.2\,\text{kN m}$$

In the calculation the factor S_2 has been kept constant (Fig. 12.3), which means the design will be a bit conservative. However, the reader can vary the S_2 factor as given in Fig. 12.3 taken from Table 3 (CP 3) which means the wind speed will be variable depending on the height of the building.

12.5.3 Assumed section of wall resisting the wind moment

The flange which acts together with the web of I-section is the lesser of

- 12 times thickness of flange + thickness of web
- centre line to centre line of walls
- one-third of span

(a) Wall A

For wall A (Fig. 12.4), neglecting the outer skin of the cavity wall flange, the second moment of area is

$$I_A = 2 \times \left(\frac{(0.1025)^3 \times 1.34}{12} + 0.1025 \times 1.34 \times (2.07)^2\right)$$

$$+ \frac{(4.045)^3 \times 0.1025}{12}$$

$$= 1.169 \times 0.565 = 1.734\,\text{m}^4$$

(b) Wall B

The flange width which acts with channel section has been assumed as half of the I-section. For wall B (Fig. 12.5), neglecting the outer skin of the cavity wall flange,

$$I_B = 2 \times \left(0.67 + \frac{(0.1025)^3}{12} + 0.1025 \times 0.67 \times (2.07)^2\right)$$

$$+ 2 \times 0.1025 \times \frac{(4.045)^3}{12}$$

$$= 0.571 \times 1.13 = 1.7\,\text{m}^4$$

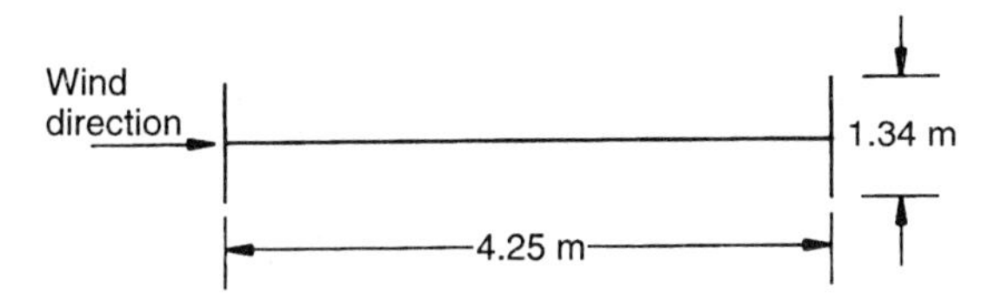

Fig. 12.4 Dimensions for wall A.

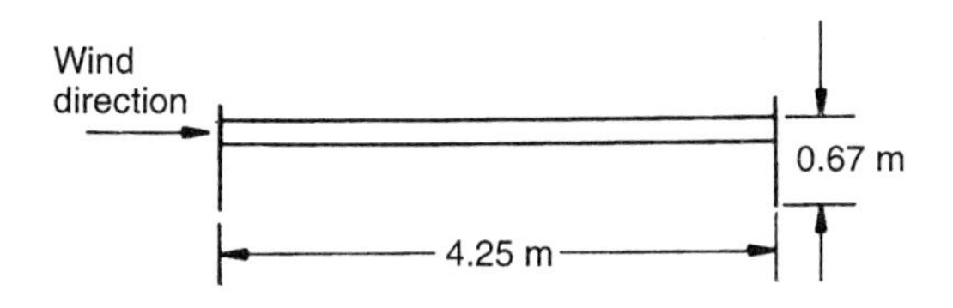

Fig. 12.5 Dimensions for wall B.

Total second moment of area for the building

$$\Sigma I = 12 I_A + 4 I_B$$
$$= 12 \times 1.734 + 4 \times 1.7 = 27.61\ \text{m}^4$$

Moment carried by wall A

$$M_A = \text{total moment} \times I_A/\Sigma I = (1.734/27.61)\,M$$
$$= 0.06266\,M$$

and moment carried by wall B

$$M_B = (1.7/27.61)M = 0.0616\,M$$

Similarly, shear force carried by wall A

$$SF_A = \text{total force} \times I_A/\Sigma I = 0.06266\,F$$

and shear force carried by wall B

$$SF_B = \text{total force} \times I_A/\Sigma I = 0.0616\,F$$

The calculated values of the *SF* are given in Table 12.3.

12.6 DESIGN LOAD

12.6.1 Load combination for ultimate limit state, wall A: clause 22, BS 5628

(a) Sixth floor

(i) Dead and imposed loads

dead + imposed = $1.4\,G_k + 1.6\,Q_k$

Table 12.3 Distribution of bending moment stresses and shear force in walls

Just above floor level		*Wall A*		*Wall B*	
		Bending stress (N/mm²)	*Shear force* (kN)	*Bending stress* (N/mm²)	*Shear force* (kN)
6th floor					
wall A	$= \frac{M_A Y}{I_A} = \frac{0.06266 \times 131.9^a}{1.734} \times \frac{2.125}{10^3}$	±0.01	5.5	±0.01	5.41
wall B	$= \frac{M_b Y}{I_b} = \frac{0.0616 \times 131.9^a}{1.7 \times 10^3} \times 2.125$				
5th floor					
wall A	$= \frac{0.06266 \times 527.6^a \times 2.125}{1.734 \times 10^3}$	±0.04	11.0	±0.04	10.83
wall B	$= \frac{0.0616 \times 527.6 \times 2.125}{1.7 \times 10^3}$				
4th floor					
wall A	$= \frac{0.06266 \times 2.125}{1.734 \times 10^3} \times 1187.2^a$				
	$= 0.0000768 \times 1187.2$	±0.09	16.5	±0.09	16.24
wall B	$= \frac{0.0616 \times 2.125}{1.7 \times 10^3} \times 1187.2^a$				
	$= 0.000077 \times 1187.2$				
3rd floor					
wall A	$= 0.768 \times 10^{-4} \times 2110.54^a$	±0.162	22.0	±0.163	21.65
wall B	$= 0.77 \times 10^{-4} \times 2110.54^a$				
2nd floor					
wall A	$= 0.768 \times 10^{-4} \times 3297.7^a$	±0.253	27.5	±0.254	27.06
wall B	$= 0.77 \times 10^{-4} \times 3297.7^a$				
1st floor					
wall A	$= 0.768 \times 10^{-4} \times 4748.71^a$	±0.365	33.0	±0.366	32.5
wall B	$= 0.77 \times 10^{-4} \times 4748.71^a$				
Ground floor					
wall A	$= 0.768 \times 10^{-4} \times 6463.72^a$	±0.496	38.50	±0.498	37.9
wall B	$= 0.77 \times 10^{-4} \times 6463.72^a$				

[a] From section 12.5.2

$$= 1.4 \times 20 + 1.6 \times 5.4$$

$$= 28 + 8.64 = 36.64\,\text{kN/m} \quad (12.1)$$

$$\text{stress} = (36.64 \times 10^3)/(102.5 \times 10^3) = 0.357\,\text{N/mm}^2 \quad (12.2)$$

(ii) Dead and wind loads

- Windward side

$$\text{dead} + \text{wind} = 0.9\,G_k + 1.4\,W_k \quad (12.3)$$

stress $= (0.9 \times 20 \times 10^3)/(102.5 \times 10^3) - 1.4 \times 0.01$

(from Table 12.3)

$$= 0.176 - 0.014 = +0.162\,\text{N/mm}^2 \quad (12.4)$$

No tension develops, hence safe.

- Leeward side

$$\text{dead} + \text{wind} = 1.4\,G_k + 1.4\,W_k \quad (12.5)$$

stress $= (28 \times 10^3)/(102.5 \times 10^3) + 0.014$ (from above)

$$= 0.273 + 0.014 = 0.287\,\text{N/mm}^2 \quad (12.6)$$

(iii) Dead, live and wind loads

$$\text{dead} + \text{live} + \text{wind} = 1.2\,G_k + 1.2\,Q_k + 1.2\,W_k \quad (12.7)$$

stress $= (0.273 \times 1.2)/1.4 + (8.64 \times 1.2)/(1.6 \times 102.5)$

$\pm 1.2 \times 0.01$ (proportionately reduced from (12.6))

$$= 0.234 + 0.0632 \times 0.012 = 0.31 \text{ or } 0.285\,\text{N/mm}^2 \quad (12.8)$$

No tension developing, hence safe.

The load combination which produces the severe condition is (12.1), and therefore, the design load $= 36.64\,\text{kN/m}$.

(b) Fifth floor

(i) Dead and imposed loads

$$\text{dead} + \text{imposed} = 1.4\,G_k + 1.6\,Q_k$$

$$= 1.4 \times 44.68 + 1.6 \times 9.72$$

$$= 62.55 + 15.55 = 78.10\,\text{kN/m} \quad (12.9)$$

$$\text{stress} = (78.10 \times 10^3)/(102.5 \times 10^3) = 0.76\,\text{N/mm}^2 \quad (12.10)$$

(ii) Dead and wind loads

- Windward side

$$\text{dead} + \text{wind} = 0.9\,G_k + 1.4\,W_k \tag{12.11}$$

$$\text{stress} = (0.9 \times 44.68 \times 10^3)/(102.5 \times 1000) - 1.4 \times 0.04$$

$$= 0.39 - 0.056 = 0.334\,\text{N/mm}^2 \tag{12.12}$$

No tension develops, hence safe.

- Leeward side

$$\text{dead} + \text{wind} = 1.4\,G_k + 1.4\,W_k \tag{12.13}$$

$$\text{stress} = (62.55 \times 10^3)/(102.5 \times 10^3) + 0.056$$

$$= 0.61 + 0.056 = 0.67\,\text{N/mm}^2 \tag{12.14}$$

(iii) Dead, live and wind loads

$$\text{dead} + \text{live} + \text{wind} = 1.2\,G_k + 1.2\,Q_k + 1.2\,W_k \tag{12.15}$$

$$\text{stress} = (0.61 \times 1.2)/1.4 + (15.55 \times 1.2)/(1.6 \times 102.5)$$

$$\pm 1.2 \times 0.04$$

$$= 0.52 + 0.11 + 0.048$$

$$= 0.68 \text{ or } 0.58\,\text{N/mm}^2 \tag{12.16}$$

No tension developing, hence safe.

Hence the load combination which produces the severe condition is (12.9) and the load is 78.10 kN/m.

(c) Fourth floor

(i) Dead and imposed loads

$$\text{dead} + \text{imposed} = 1.4\,G_k + 1.6\,Q_k$$

$$= 1.4 \times 69.36 + 1.6 \times 12.96$$

$$= 97.10 + 20.74\,\text{kN/m} \tag{12.17}$$

$$\text{stress} = (97.10 \times 10^3)/(102.5 \times 10^3) + (20.74 \times 10^3)(102.5 \times 10^3)$$

$$= 0.95 + 0.20 = 1.15\,\text{N/mm}^2 \tag{12.18}$$

(ii) Dead and wind loads

- Windward side

$$\text{dead} + \text{wind} = 0.9\,G_k + 1.4\,W_k \tag{12.19}$$

$$\text{stress} = (0.95 \times 0.9)/1.4 - 1.4 \times 0.09$$

(proportionally reduced from (12.18))

$$= 0.61 - 0.126 = 0.484\,\text{N/mm}^2 \text{ (no tension)} \quad (12.20)$$

- Leeward side

$$\text{dead} + \text{wind} = 1.4\,G_k + 1.4\,W_k \quad (12.21)$$

$$\text{stress} = 0.95 + 0.126 = 1.08\,\text{N/mm}^2 \quad (12.22)$$

(iii) Dead, imposed and wind loads

$$\text{dead} + \text{imposed} + \text{wind} = 1.2\,G_k + 1.2\,Q_k + 1.2\,W_k \quad (12.23)$$

$$\text{stress} = (0.95 \times 1.2)/1.4 + (1.2 \times 0.20)/1.6 \pm 1.2 \times 0.09$$

$$= 0.814 + 0.15 \pm 0.108$$

$$= 1.07 \text{ or } 0.856\,\text{N/mm}^2 \quad \text{(no tension)} \quad (12.24)$$

In this case also the severe loading condition appears to be (12.17).

(d) Third floor

(i) Design and imposed loads

$$\text{design} + \text{imposed} = 1.4\,G_k + 1.6\,Q_k$$

$$= 1.4 \times 94.04 + 1.6 \times 15.12$$

$$= 131.66 + 24.19 = 155.85\,\text{kN/m} \quad (12.25)$$

$$\text{stress} = (131.66 \times 10^3)/(102.5 \times 10^3)$$

$$+ (24.19 \times 10^3)/(102.5 \times 10^3)$$

$$= 1.28 + 0.24 = 1.52\,\text{N/mm}^2 \quad (12.26)$$

(ii) Dead and wind loads

- Windward side

$$\text{dead} + \text{wind} = 0.9\,G_k + 1.4\,W_k \quad (12.27)$$

$$\text{stress} = (0.9 \times 1.28)/1.4 - 1.4 \times 0.162$$

$$= 0.823 - 0.227 = 0.596\,\text{N/mm}^2 \quad \text{(no tension)} \quad (12.28)$$

- Leeward side

$$\text{dead} + \text{wind} = 1.4\,G_k + 1.4\,W_k \quad (12.29)$$

$$\text{stress} = 1.28 + 0.227 = 1.51\,\text{N/mm}^2 \quad (12.30)$$

(iii) Dead, live and wind loads

$$\text{dead} + \text{live} + \text{wind} = 1.2\,G_k + 1.2\,Q_k + 1.2\,W_k \qquad (12.31)$$

$$\begin{aligned} \text{stress} &= (1.28 \times 1.2)/1.4 + (0.24 \times 1.2)/1.6 \pm 1.2 \times 0.162 \\ &= 1.097 + 0.18 \pm 0.194 \\ &= 1.47 \text{ or } 1.08\,\text{N/mm}^2 \text{ (no tension develops)} \qquad (12.32) \end{aligned}$$

The critical load combination is (12.25) and the load is 155.85 kN/m.

(e) Second floor

(i) Design and imposed loads

$$\begin{aligned} \text{design} + \text{imposed} &= 1.4\,G_k + 1.6\,Q_k \\ &= 1.4 \times 118.72 + 1.6 \times 16.2 \\ &= 166.2 + 25.9 = 192.10\,\text{kN/m} \qquad (12.33) \end{aligned}$$

$$\begin{aligned} \text{stress} &= (1.4 \times 118.72 \times 10^3)/(102.5 \times 10^3) \\ &\quad + (1.6 \times 16.2 \times 10^3)/(102.5 \times 10^3) \\ &= 1.62 + 0.25 = 1.87\,\text{N/mm}^2 \qquad (12.34) \end{aligned}$$

(ii) Dead and wind loads

- Windward side

$$\text{dead} + \text{wind} = 0.9\,G_k + 1.4\,W_k \qquad (12.35)$$

$$\begin{aligned} \text{stress} &= (0.9 \times 1.62)/1.4 - (1.4 \times 0.253) \\ &= 1.04 - 0.35 = 0.69\,\text{N/mm}^2 \quad \text{(no tension)} \qquad (12.36) \end{aligned}$$

- Leeward side

$$\text{dead} + \text{wind} = 1.4\,G_k + 1.4\,W_k \qquad (12.37)$$

$$\text{stress} = 1.62 + 1.4 \times 0.253 = 1.97\,\text{N/mm}^2 \qquad (12.38)$$

(iii) Dead, imposed and wind loads

$$\text{dead} + \text{imposed} + \text{wind} = 1.2\,G_k + 1.2\,Q_k + 1.2\,W_k \qquad (12.39)$$

$$\begin{aligned} \text{stress} &= (1.62 \times 1.2)/1.4 + (0.25 \times 1.2)/1.6 + 1.2 \times 0.253 \\ &= 1.39 + 0.19 \pm 0.30 \\ &= 1.88 \text{ or } 1.28\,\text{N/mm}^2 \text{ (no tension)} \qquad (12.40) \end{aligned}$$

The critical load combination which produces the severe condition is (12.37) and the design load therefore is $(1.97 \times 102.5 \times 10^3)/10^3 = 202\,\text{kN/m}$.

(f) First floor

(i) Dead and imposed loads

$$\text{dead} + \text{imposed} = 1.4\,G_k + 1.6\,Q_k$$
$$= 1.4 \times 143.40 + 1.6 \times 19.44$$
$$= 200.76 + 31.10 = 231.86\,\text{kN/m} \quad (12.41)$$
$$\text{stress} = (200.76 \times 10^3)/(102.5 \times 10^3) + (31.10 \times 10^3)/(102.5 \times 10^3)$$
$$= 1.96 + 0.30 = 2.26\,\text{N/mm}^2 \quad (12.42)$$

(ii) Dead and wind loads

- Windward side

$$\text{dead} + \text{wind} = 0.9\,G_k + 1.4\,W_k \quad (12.43)$$
$$\text{stress} = (1.96 \times 0.9)/1.4 - 1.4 \times 0.365$$
$$= 1.26 - 0.51 = 0.75\,\text{N/mm}^2 \text{ (no tension)} \quad (12.44)$$

- Leeward side

$$\text{dead} + \text{wind} = 1.4\,G_k + 1.4\,W_k \quad (12.45)$$
$$\text{stress} = 1.96 + 0.51 = 2.47\,\text{N/mm}^2 \quad (12.46)$$

(iii) Dead, imposed and wind loads

$$\text{dead} + \text{imposed} + \text{wind} = 1.2\,G_k + 1.2\,Q_k + 1.2\,W_k \quad (12.47)$$
$$\text{stress} = (1.2 \times 1.96)/1.4 + (1.2 \times 0.3)/1.6 \pm (1.2 \times 0.51)/1.4$$
$$= 1.68 + 0.225 \pm 0.437$$
$$= 2.34 \text{ or } 1.47\,\text{N/mm}^2 \text{ (no tension develops)} \quad (12.48)$$

The critical load combination is (12.45) and the design load for this floor is $(2.47 \times 102.5 \times 10^3)/10^3 = 253.18\,\text{kN/m}$.

(g) Ground floor

(i) Dead and imposed loads

$$\text{dead} + \text{imposed} = 1.4\,G_k + 1.6\,Q_k \quad (12.49)$$
$$\text{stress} = (1.4 \times 168.08)/102.5 + (1.6 \times 22.68)/102.5$$
$$= 2.3 + 0.354 = 2.654\,\text{N/mm}^2 \quad (12.50)$$

(ii) Dead and wind loads

- Windward side

$$\text{dead} + \text{wind} = 0.9\,G_k + 1.4\,W_k \quad (12.51)$$

$$\text{stress} = (2.3 \times 0.9)/1.4 - 1.4 \times 0.496$$

(proportionally reduced from (12.50))

$$= 1.48 - 0.69 = 0.78\,\text{N/mm}^2 \quad \text{(no tension)} \quad (12.52)$$

- Leeward side

$$\text{dead} + \text{wind} = 1.4\,G_k + 1.4\,W_k \quad (12.53)$$

$$\text{stress} = 2.3 + 0.69 = 2.99\,\text{N/mm}^2 \quad (12.54)$$

(iii) Dead, imposed and wind loads

$$\text{dead} + \text{imposed} + \text{wind} = 1.2\,G_k + 1.2\,Q_k + 1.2\,W_k \quad (12.55)$$

$$\text{stress} = (2.3 \times 1.2)/1.4 + (1.2 \times 0.354)/1.6 \pm 1.2 \times 0.496$$

$$= 1.97 + 0.266 \pm 0.595$$

$$= 2.83 \text{ or } 1.64\,\text{N/mm}^2 \quad \text{(no tension develops)} \quad (12.56)$$

The load combination (12.53) produces the severe condition and hence the design load is $(2.99 \times 102.5 \times 10^3)/10^3 = 306.48\,\text{kN/m}$.

Note that from section 12.5.2 the total wind force

$$F = C_f q A_c$$

$$= (1.1 \times 1269)/10^3 \times 21 \times 21$$

$$= 615.6\,\text{kN}$$

and

$$615.6 \times \gamma_f > 0.015\,G_k \quad (G_k \text{ from section 12.4.3})$$

or

$$615.6 \times 1.4 = 861.84\,\text{kN} > 0.015 \times 17643 = 264.65\,\text{kN}$$

Hence in the load combination $0.015\,G_k$ has not been considered. This is true for all other floors also.

12.6.2 Selection of brick and mortar combinations for wall A: BS 5628

Design vertical load resistance of wall is $(\beta t f_k)/\gamma_m$ (clause 32.2.1), eccentricity $e = 0$, $SR = \frac{3}{4} \times (2.85 \times 10^3)/102.5 = 20.85$. Hence $\beta = 0.67$ (Table 7 of BS 5628), $\gamma_m = 3.5$ (see section 12.3). The design loads from the previous subsection and the characteristic strengths are shown in Table 12.4 along with the suitable brick/mortar combinations.

Check for shear stress: design characteristic shear $f_v = \gamma_f \gamma_{mv}$ (shear force/area) $< 0.35 + 0.6 g_A$ (clause 25), $\gamma_f = 1.4$ and $\gamma_{mv} = 2.5$ (12.3). The value of shear force is taken from Table 12.3. For the sixth floor

$$\text{design characteristic shear stress} = \frac{1.4 \times 2.5 \times 5.5 \times 10^3}{102.5 \times 4250} = 0.044\,\text{N/mm}^2$$

$$< 0.35 + \frac{0.6 \times 20 \times 0.9 \times 10^3}{102.5 \times 1000} = 0.45 \quad \text{(safe)}$$

For the ground floor

$$\text{design characteristic shear stress} = \frac{1.4 \times 2.5 \times 38.5 \times 10^3}{102.5 \times 4250} = 0.31\,\text{N/mm}^2$$

$$< 0.35 + \frac{0.6 \times 168.08 \times 0.9 \times 10^3}{102.5 \times 1000} = 1.35\,\text{N/mm}^2$$

There is no need to check at any other level, since shear is not a problem for this type of structure.

The BS 5628 recommends g_A as the design vertical load per unit area of wall cross-section due to vertical load calculated from the appropriate loading condition specified in clause 22. The critical condition of shear will be with no imposed load just after and during the construction.

12.6.3 Load combination, wall B

The design principle has been covered in great detail for wall A; hence for wall B this will be limited to the ground floor level to explain further salient points.

Inner leaf wall B – ground floor level

(i) Dead and imposed loads

$$\text{dead} + \text{imposed} = 1.4\,G_k + 1.6\,Q_k \quad (G_k \text{ and } Q_k \text{ from Table 12.2})$$

$$\text{stress} = (1.4 \times 91.9 \times 10^3)/(102.5 \times 10^3)$$

$$+ (1.6 \times 8.51 \times 10^3)/(102.5 \times 10^3)$$

$$= 1.26 + 0.13 = 1.39\,\text{N/mm}^2$$

Table 12.4 Design load and characteristic brickwork strength

Floor	*Design load/m (section 12.6.1)*	*Design characteristic strength* f_k (N/mm²) $= \frac{\text{design load} \times \gamma_m}{\beta t}$	f_k *from table 2 and clause 23.1.2* (N/mm²)
6th	36.64	1.87	20 N/mm² brick in 1:1:6 mortar $f_k = 1.15 \times 5.8 = 6.67$ N/mm²
5th	78.10	3.98	20 N/mm² brick in 1:1:6 mortar $f_k = 1.15 \times 5.8 = 6.67$ N/mm²
4th	117.14	6.0	20 N/mm² brick in 1:1:6 mortar $f_k = 1.15 \times 5.8 = 6.67$ N/mm²
3rd	155.85	7.94	20 N/mm² brick in 1:$\frac{1}{4}$:3 mortar $f_k = 1.15 \times 7 = 8.51$ N/mm²
2nd	202.0	10.29	35 N/mm² brick in 1:$\frac{1}{4}$:3 mortar $f_k = 1.15 \times 11.4 = 13.11$ N/mm²
1st	253.18	12.9	35 N/mm² brick in 1:$\frac{1}{4}$:3 mortar $f_k = 1.15 \times 11.4 = 13.11$ N/mm²
GF	306.48	15.62	50 N/mm² brick in 1:$\frac{1}{4}$:3 mortar $f_k = 1.15 \times 15 = 17.25$ N/mm²

(ii) Dead and wind loads

- Windward side

$$\text{dead} + \text{wind} = 0.9\,G_k + 1.4\,W_k$$

$$\text{stress} = (0.9 \times 1.26)/1.4 - 1.4 \times 0.498 \text{ (from Table 12.3);}$$

(proportionately reduced from above)

$$= 0.81 - 0.70 = 0.11\,\text{N/mm}^2 \quad \text{(no tension develops)}$$

- Leeward side

$$\text{dead} + \text{wind} = 1.4\,G_k + 1.4\,W_k$$

$$\text{stress} = 1.26 + 1.4 \times 0.498 = 1.96\,\text{N/mm}^2$$

(iii) Dead, imposed and wind loads

$$\text{dead} + \text{imposed} + \text{wind} = 1.2\,G_k + 1.2\,Q_k + 1.2\,W_k$$

$$\text{stress} = (1.26 \times 1.2)/1.4 + (0.13 \times 1.2)/1.6 \pm 1.2 \times 0.498$$

$$= 1.08 + 0.098 \pm 0.98$$

$$= 1.78 \text{ or } 0.58\,\text{N/mm}^2 \quad \text{(no tension develops)}$$

The worst combination for this wall just above ground level also is dead + wind, and the design load is $(1.96 \times 102.5 \times 10^3)/10^3 = 201\,\text{kN/m}$.

12.6.4 Selection of brick and mortar for inner leaf of wall B

The design vertical load resistance of the wall is $(\beta t f_k)/\gamma_m$ (clause 32.2.1). The value of β depends on the eccentricity of loading; hence the value of e needs to be evaluated before design can be completed.

12.6.5 Calculation of eccentricity

The worst combination of loading for obtaining the value of e at top of the wall is shown in Fig. 12.6. Axial load

$$P = (0.9 \times 78.54 + 1.6 \times 7.29) \quad (G_k \text{ and } Q_k \text{ from Table 12.2})$$

$$= (70.69 + 11.66) = 82.35\,\text{kN/m}$$

First floor load

$$P_1 = (1.4 \times 6.48 + 1.6 \times 2.025) \quad \text{(see Table 12.2)}$$

$$= 12.31\,\text{kN/m}$$

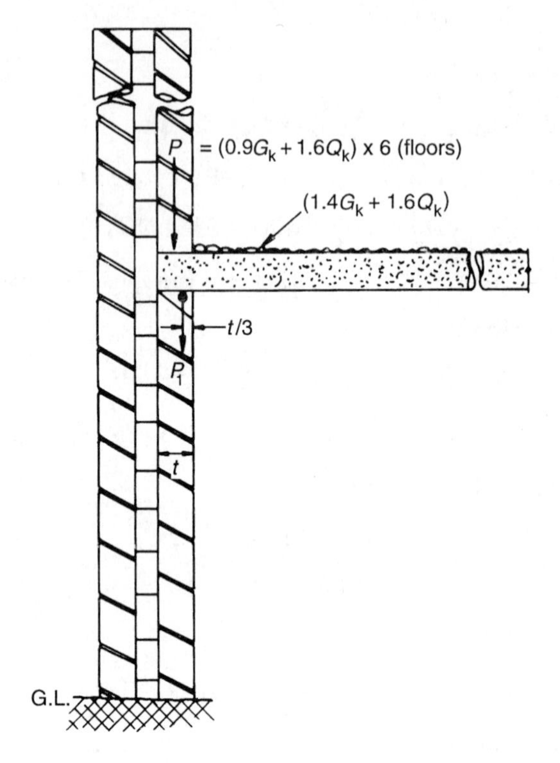

Fig. 12.6 Load combination for calculating the eccentricity.

Eccentricity

$$e = \frac{P_1 t}{6(P + P_1)} = \frac{12.31\,t}{6(82.35 + 12.31)} = 0.0217t = 2.22\,\text{mm}$$

(a) Wind blowing north–south direction

A part of the panel B will be subjected to suction, if the wind is blowing in N–S direction. Then

$V_s = VS_1S_2S_3 = 50 \times 1 \times 1 \times 0.64$ (ground roughness category A, CP 3, Chapter V: Part 2)

$= 32\,\text{m/s}$

Note that the localized effect is considered here, hence S_2 for Category A is being used. Also

$$q = 0.613 \times (32)^2 = 627.8\,\text{N/mm}^2$$

$$\text{BM at centre of the panel} = 627.8 \times (Cp_e + CP_i)h^2 \times 0.104 \times 1.4$$

$$= 627.8 \times (1.1 + 0.2) \times (2.85)^2 \times 0.104 \times 1.4$$

$$= 964.6\,\text{Nm/m} \quad (Cp_e \text{ and } Cp_i \text{ from CP 3, Chapter V: Part 2})$$

(BM coefficient for four-sided simply supported panel is 0.104; table 3.1, BS 8110)

$$\text{BM/leaf} = 964.6/2 = 482.3\,\text{Nm/m}$$

(since both leaves are of same stiffness)

$$e_{\text{centre}} = (482.3 \times 10^3)/(94.66 \times 10^3) = 5.1\text{ mm}$$

where

$$P + P_1 = 94.66\,\text{kN/m}$$

Resultant

$$e_{cc} = (2.22/2) + 5.1 = 6.21\,\text{mm} = 0.06t$$

(b) Wind blowing west–east direction

The panel B is not only subjected to dead and imposed loads, but also subjected to wind loading from west to east direction. Then

$$\text{BM at the centre} = 0.104 \times q h^2 \times \gamma_f \quad (q \text{ from section 12.5.2})$$

$$= 0.104 \times 1.4 \times 1269 \times 1.1 \times (2.85)^2$$

(considering the loading from overall stability)

$$= 1650.84\,\text{Nm/m}$$

$$\text{BM/leaf (as before)} = 1650.85/2 = 825.42\,\text{Nm/m}$$

$$e_{cc} \text{ at the centre} = (825.42 \times 10^3)/(94.66 \times 10^3) = 8.72\,\text{mm}$$

Therefore resultant

$$e = 8.72 - (2.22/2)$$

$$= 7.61\,\text{mm} = 0.074t$$

(the bending moment induced due to wind loading acts against those due to the vertical load).

Since resultant eccentricity of case (b) is greater than case (a), case (b) eccentricity is considered in the design.

12.6.6 Calculation of characteristic compressive stress f_k for wall B (inner leaf)

$$\text{design load} = (\beta t f_k)/\gamma_m \quad \text{(clause 32.2.1, BS 5628)}$$

$$\text{slenderness ratio} = (\tfrac{3}{4} \times 2.85 \times 10^3)/[\tfrac{2}{3}(102.5 + 102.5)] = 15.6$$

and

$$e_R = 0.074t$$

Therefore

$$\beta \text{ (from table 7)} = 0.81 \text{ from linear interpolation.}$$

Therefore

$$0.81 \times 102.5 f_k = 3.5 \times 201$$

$$f_k = 8.47\,\text{N/mm}^2$$

Use $20\,\text{N/mm}^2$ brick in $1{:}\tfrac{1}{4}{:}3$ mortar,

$$f_k = 7.4 \times 1.15 = 8.51\,\text{N/mm}^2 > 8.47\,\text{N/mm}^2 \quad \text{(safe),}$$

Check for shear:

$$\text{design characteristic shear stress} = \gamma_{mv} \times \frac{\text{shear force}}{\text{area}} \gamma_f$$

$$= \frac{2.5 \times 37.9 \times 10^3}{2 \times 102.5 \times 4250} \times 1.4 = 0.15\,\text{N/mm}^2$$

$$< 0.35 + \frac{0.9 \times 91.90 \times 10^3 \times 0.6}{1000 \times 102.5}$$

$$= 0.834\,\text{N/mm}^2 \quad \text{(safe),}$$

12.6.7 Design of the outer leaf of the cavity wall B in GF

Load combination:

- Windward side

$$\text{dead} + \text{imposed} = 0.9\,G_k + 1.2\,W_k \quad \text{(300 mm projection of roof)}$$

Note: $\gamma_f = 1.2$ is used as per clause 22.

$$\text{stress} = \frac{0.9\,(7 \times 7.26 + 0.30 \times 3.5)\,10^3}{102.5 \times 1000} - 1.2 \times 0.498$$

$$= 0.455 - 0.598 = -0.14\,\text{N/mm}^2$$

- Leeward side

$$\text{dead} + \text{imposed} = 1.4G_k + 1.2W_k$$

$$\text{stress} = 1.4 \times 0.455 + 0.598 = 1.23\,\text{N/mm}^2$$

The design is similar to the inner leaf and will not be considered any further. The slight tension which is developing is of no consequence, since 6 to 10% of the dead and imposed load will be transferred to the outer leaf even in cases where the slab is supported on the inner skin. The bending stress caused by the wind will be smaller if S_2 factor is assumed variable as explained in section 12.5.2: the staircase and lift well will also provide the stability against the wind which has been neglected. However, any facing brick having water absorption between 7 and 12% in $1{:}\frac{1}{4}{:}3$ mortar may be used, provided that it satisfies the lateral load design. The grade of mortar is kept the same as for the inner leaf.

Characteristic flexural strength:

$$f_{ky} = 0.14 \times 3.5 \quad (\gamma_m = 3.5)$$

$$= 0.49\,\text{N/mm}^2 < 0.5\,\text{N/mm}^2 \quad \text{(Table 3)}$$

Design characteristic shear as in inner leaf:

$$0.15 < 0.35 + \frac{0.9 \times (7 \times 7.26 \times 0.3 \times 3.5) \times 10^3}{102.5 \times 10^3} \times 0.6 = 0.65\,\text{N/mm}^2\ \text{(safe)}$$

Instead of the conventional design calculations described in this chapter a more sophisticated analysis of the structure is possible by idealizing it as a frame with vertical loading as shown in Fig. 12.7. Similarly, the structure can be idealized and replaced by a two-dimensional frame (Fig. 12.8) and analysed as discussed in Chapter 6 for wind loading.

12.7 DESIGN CALCULATION ACCORDING TO EC6 PART 1-1 (ENV 1996-1: 1995)

To demonstrate the principle of design according to EC6, the wall A in the ground floor will be redesigned. The dead and live loading is taken as calculated before and as in Table 12.1. The bending moments and shear forces due to wind loading are given in Table 12.3. The category of manufacturing and execution controls are assumed to be II and C respectively; thus $\gamma_m = 3$ as given in Table 4.6.

Load combination for ultimate limit state:

$$\text{permanent} + \text{variable} = 1.35\,G_{kj} + 1.5\,Q_{ki}$$

$$\text{stress} = (1.35 \times 168.08 + 1.5 \times 22.68)/102.5$$

$$= 2.2 + 0.33 = 2.53\,\text{N/mm}^2$$

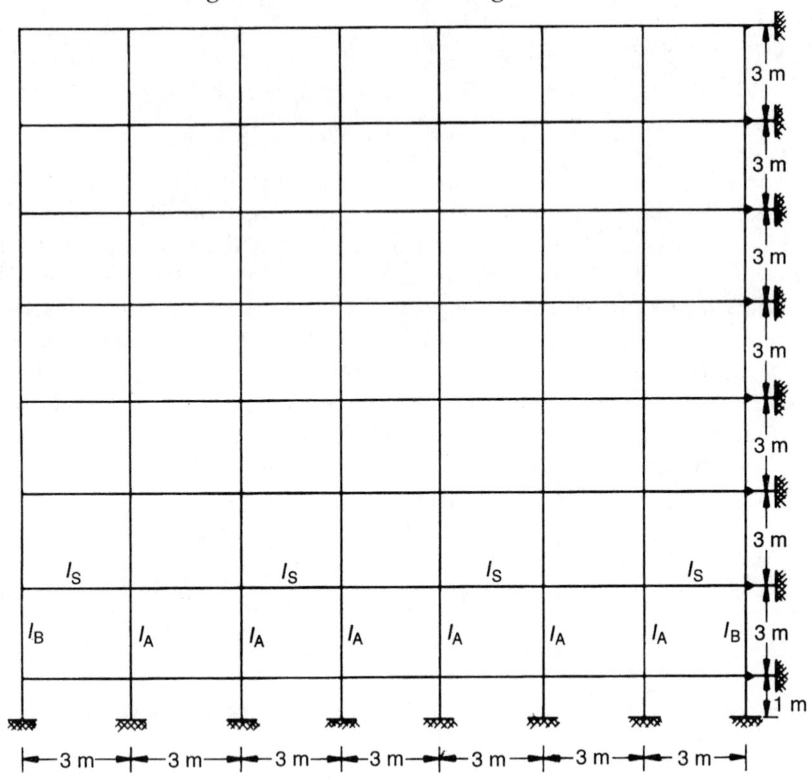

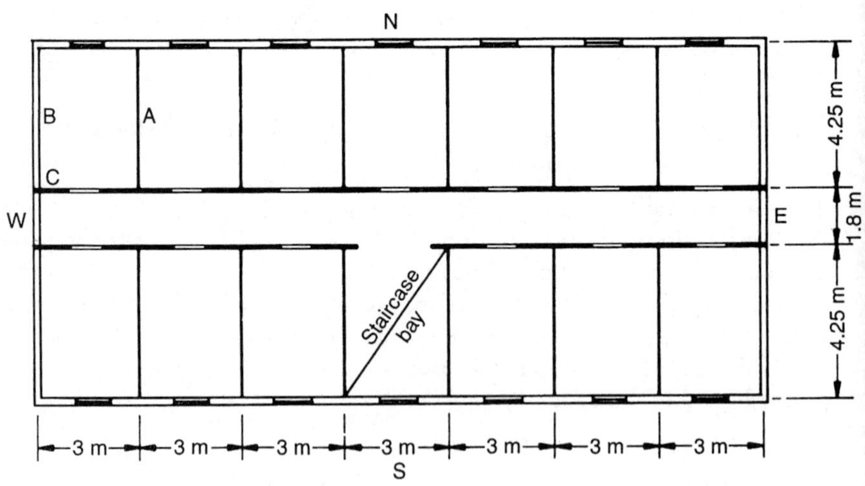

Fig. 12.7 Idealized structure for vertical load design.

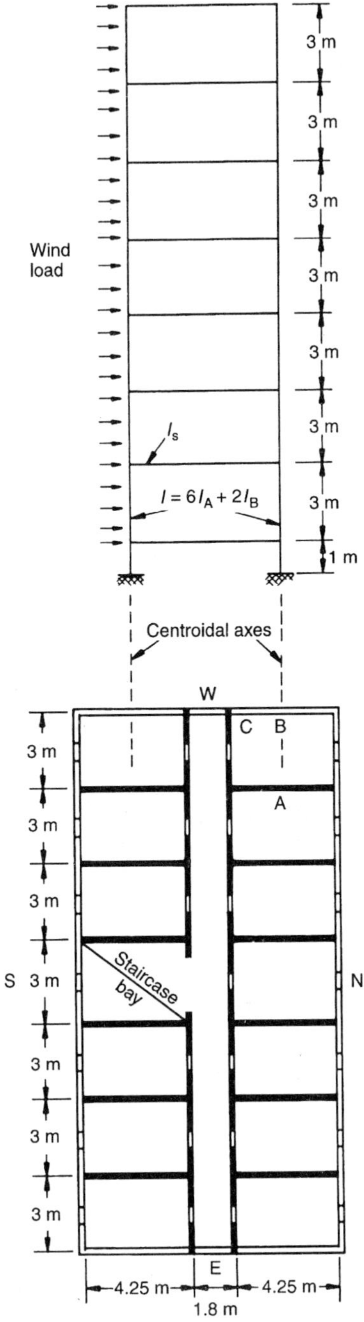

Fig. 12.8 Idealized structure for wind load design.

- Windward side

$$\text{permanent} + \text{variable} = 1G_{kj} - 1.5\,W_{ki}$$

$$\text{stress} = \frac{1 \times 168.08}{102.5} - 1.5 \times 4.96 = 1.64 - 0.744$$

$$= 0.896\,\text{N/mm}^2 \quad \text{(no tension)}$$

- Leeward side

$$\text{permanent} + \text{variable} = 1.35\,G_{kj} + 1.5\,W_{ki}$$

$$\text{stress} = 2.2 + 0.744 = 2.944\,\text{N/mm}^2$$

$$\text{permanent} + \text{all variable actions} = 1.35\,G_{kj} + 1.35\,Q_{ki} \pm 1.35\,W_{ki}$$

$$\text{stress} = \frac{1.35 \times 168.08}{102.5} + \frac{1.35 \times 22.68}{102.5}$$

$$\pm 1.35 \times 0.496$$

$$= 2.20 + 0.3 \pm 0.67$$

$$= 3.17 \text{ or } 1.83\,\text{N/mm}^2 \quad \text{(no tension)}$$

Hence, most unfavourable action is $1.35\,G_{kj} + 1.35\,Q_{ki} + 1.35\,W_{ki}$ and the design load $= 3.17 \times 102.5 \times 10^3/10^3 = 324.9\,\text{kN/m}$.

12.7.1 Selection of brick and mortar combination for wall A: according to EC6

Design vertical load resistance of wall $N_{Rd} = \phi_{i,m}\,t f_k/\gamma_m$, where $\phi_{i,m}$ depends on eccentricity and slenderness ratio $SR = \frac{3}{4} \times 2.85/102.5 \times 10^3 = 20.85$.

12.7.2 Calculation of eccentricity

Figure 12.9 shows the worst combination of loading for obtaining the value of eccentricity. Axial load

$$P = 1 \times 143.40 + 1.5 \times 19.44$$

$$= 172.56\,\text{kN}$$

$$P_1 = 1.35 \times 4.8 \times 1.2 \times 1.5 + 1.5 \times 2.7$$

$$= 11.66 + 4.05$$

$$= 15.71\,\text{kN}$$

$$P_2 = 1.35 \times G_k = 11.66\,\text{kN}$$

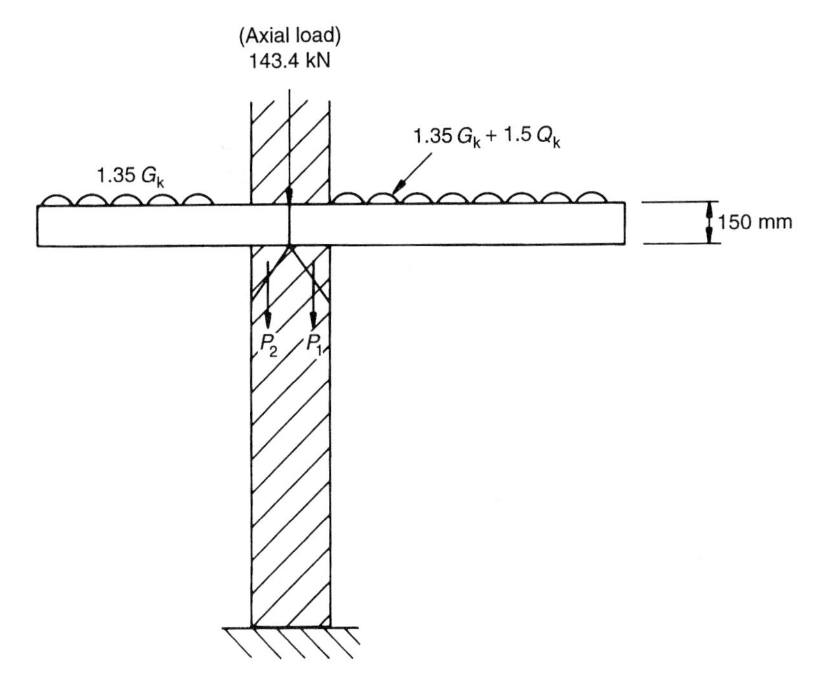

Fig. 12.9 Calculation of eccentricity of the loading (not to scale).

$$e_{cc} = 11.66 \times 17.08 + \frac{172.56 \times 51.25}{199.93} + 15.71 \times 85.45$$

$$= \frac{199.15 + 8897.19 + 1342.42}{199.93}$$

$$= \frac{10\,438.76}{199.93} = 52.21 \text{ mm}$$

eccentricity $= 52.21 - 51.25 = 0.96\,\text{mm}$

Total e_{cc}

$$e_i = M_i/N_i + e_2$$

$$= 0.96 + \frac{0.75 \times 2.850 \times 10^3}{450} = 0.96 + 4.75$$

$$= 5.71 = 0.056\,t$$

$$\phi_i = 1 - \frac{2\,e_i}{t}$$

$$= 1 - \frac{2 \times 0.056\,t}{t} = 0.88$$

$$f_k = \frac{324.9 \times 3}{0.59 \times 102.5} = 16.12\,\text{N/mm}^2 < 1.15 \times 15 = 17.25\,\text{N/mm}^2$$

EC6 allows the use of the value from the national code. Hence $50\,\text{N/mm}^2$ brick in $1{:}\frac{1}{4}{:}3$ mortar will be sufficient.

In the absence of test data a formula as given below is suggested for use:

$$f_k = k\delta f_b^{0.65} f_m^{0.25}\,\text{N/mm}^2$$

$$= 0.5 \times 0.85 \times f_b^{0.25} \times 16^{0.25}$$

or

$$f_b^{0.65} = \frac{16.12}{0.5 \times 0.85 \times 2} = 19$$

$$f_b = 100\,\text{N/mm}^2$$

Therefore, $100\,\text{N/mm}^2$ bricks are required which is much higher than the previous case. It would be better and economical to do tests on prisms to obtain the characteristic strength.

For the ground floor

$$\text{design characteristic shear stress } f_{\gamma k} = \gamma_Q \gamma_m \frac{\text{shear force}}{\text{area}} \leqslant f_{\gamma k0} + 0.4\sigma_\delta$$

$$= \frac{1.5 \times 3.5 \times 38.5 \times 10^3}{102.5 \times 4250} \leqslant 0.3$$

$$+ \frac{0.4 \times 1 \times 168.08 \times 10^3}{102.5 \times 1000}$$

$$= 0.464\text{N/mm}^2 < 0.96\text{N/mm}^2 \text{ (safe)}$$

γ_G has been taken as 1 for favourable effect.

The allowable shear due to precompression in BS 5628 is higher than in the Eurocode, but it does not make much difference to the design.

12.8 DESIGN OF PANEL FOR LATERAL LOADING: BS 5628 (LIMIT STATE)

To explain the principle of the design only panel B between sixth floor and roof will be considered. The low precompression on the inner leaf is ignored in this design. Assume:

- Inner leaf 102.5 mm brickwork in 1:1:6 mortar
- Outer leaf 102.5 mm brickwork with facing brick in 1:1:6 mortar
- Boundary conditions: two sides simply supported and two sides fixed as shown in Fig. 12.10.

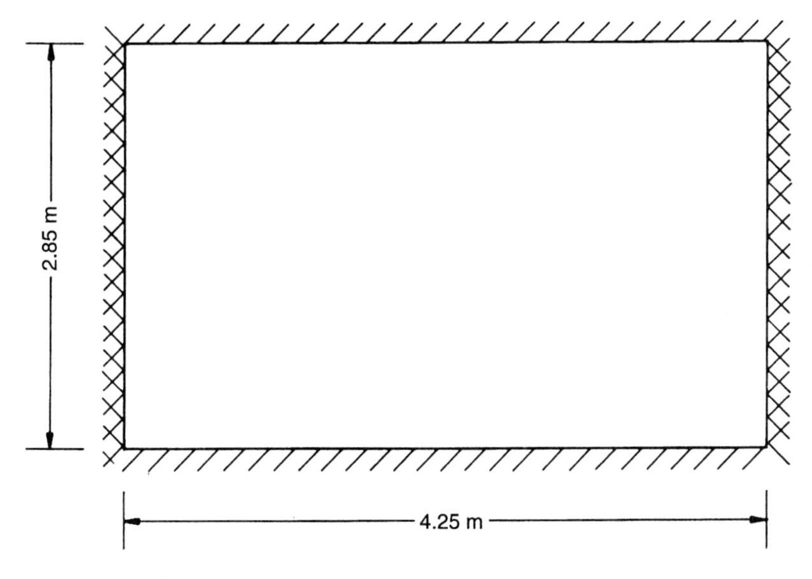

Fig. 12.10 Panel, simply supported top and bottom and fixed at its vertical edges.

12.8.1 Limiting dimension: clause 36.3, BS 5628: case B

The dimensions $h \times 1$ of panels supported on four edges should be equal to or less than 2025 $(t_{ef})^2$:

$$\text{area} = 2.85 \times 4.25 = 12.12\,\text{m}^2$$

$$t_{ef} = \tfrac{2}{3} \times 205 = 137\,\text{mm}$$

$$2025 \times (137)^2/10^6 = 38\,\text{m}^2 > 12.12\,\text{m}^2 \quad \text{(satisfactory)}$$

$$50 \times t_{ef} = 6.85 > 4.25 \quad \text{(satisfactory)}$$

12.8.2 Characteristic wind load W_k

The corner panel is subjected to local wind suctions. From CP 3, Chapter V, total coefficient of wind pressure,

$$C_p - C_{p1} = -1.1 - (+0.2) = (-)1.3$$

The design wind velocity

$$V_s = V S_1 S_2 S_3$$

where $S_1 = S_3 = 1$.

Using ground roughness Category (3), Class A, and height of the building = 21 m, therefore

$$S_3 = 0.956$$

Therefore

$$V_s = 50 \times 1 \times 1 \times 0.956 = 47.8\,\text{m/s}$$

and

$$\text{dynamic wind pressure} = [0.613 \times (47.8)^2]/10^3 = 1.4\,\text{kN/m}^2$$

Now

$$W_k = 1.4 \times 1.3 = 1.82\,\text{kN/m}^2 \quad \text{(suction)}$$

$$\text{design moment in panel} = \alpha W_k \gamma_f L^2 \quad \text{(clause 36.4.1)}$$

$$\text{aspect ratio of panel} = 2.85/4.25 = 0.67$$

therefore

$$\alpha = 0.032 \quad \text{(table 9)}$$

$$\mu = 0.33 \quad \text{(from table 3)}$$

$$\text{design moment} = 0.032 \times 1.82 \times 1.4 \times (4.25)^2 = 1.47\,\text{kN m/m}$$

Note that γ_f is taken as 1.4 since inner leaf is an important loadbearing element. The designer may, however, use $\gamma_f = 1.2$ in other circumstances.

$$\text{design moment/leaf} = 1.47/2 = 0.736\,\text{kNm/m}$$

(since both leaves are of equal stiffness)

$$\text{design moment of resistance} = f_{kx} Z/\gamma_m \quad (\text{where } \gamma_m = 3.5)$$

$$= f_{kx}(1000 \times 102.5^2)/(6 \times 3.5)$$

$$= 500\,298 f_{kx}\,\text{mm}^3/\text{m}$$

Therefore

$$f_{kx} = (0.736 \times 10^6)/(500\,298) = 1.47\,\text{N/mm}^2 < 1.5\,\text{N/mm}^2$$

Use bricks having water absorption less than 7% in 1:1:6 mortar.

12.9 DESIGN FOR ACCIDENTAL DAMAGE

12.9.1 Introduction

The building which has been designed earlier in this chapter falls in Category 2 (table 12, BS 5628) and hence the additional recommendation of clause 37 to limit the extent of accidental damage must be met over and above the recommendations in clause 20.2 for the preservation of structural integrity.

Three options are given in the code in Table 12. Before these options are discussed it would be proper to consider whether the walls A and B in the ground floor, carrying heaviest precompression, can be designated as protected elements.

12.9.2 Protected wall

A protected wall must be capable of resisting $34\,\text{kN/m}^2$ from any direction. Let us examine wall A first.

(a) Wall A

$$\text{Load combination} = 0.95\,G_k + 0.35\,Q_k + 0.35\,W_k \quad \text{(clause 22)}$$

G_k = the load just below the first floor. So

$$\text{axial stress} = \frac{10^3[0.95 \times (168.08 - 7.4) + 0.35 \times 22.68]}{102.5 \times 1000} \pm 0.35 \times 0.365$$

(see tables 12.1 and 12.3)

$$= 1.57 \pm 0.1277$$

$$= 1.442 \text{ or } 1.70\,\text{kN/mm}^2 > 0.1\,\text{N/mm}^2$$

Therefore

$$n = (1.442 \times 102.5 \times 1000)/1000 = 147.8\,\text{kN/m}$$

(b) Lateral strength of wall with two returns

$$q_{\text{lat}} = k \times 8_{tn}/h^2\gamma_m = k \times 7.6_{tn}/h^2 \quad \text{(clause 36.8 and Table 10)}$$

$$1/h = 4.25/2.85 = 1.49$$

hence $k = 2.265$. (Note that in clause 37.1.1 a factor of 7.6, which is equal to 8/1.05, has now been suggested.)

$$q_{\text{lat}} = (2.265 \times 8 \times 102.5 \times 147.8)/[(2.85)^2 \times 1.05]$$

$$= 33.8\,\text{kN/m}^2 < 34\,\text{kN/m}^2$$

Hence this wall cannot strictly be classified as a protected member.

Since wall A, carrying a higher precompression, just fails to resist $34\,\text{kN/m}^2$ pressure, wall B, with a lower precompression, obviously would not meet the requirement for a protected member.

Further, for both walls

$$h/t = (2.85 \times 10^3)/102.5 = 27.8 > 25$$

Neither wall A nor B can resist $34\,kN/m^2$. Even if they did, they do not fulfil the requirement of clause 36.8 that

$$h/t \leqslant 25$$

It may be commented that the basis of this provision in the code is obscure and conflicts with the results of tests on laterally loaded walls. Other options therefore need to be considered in designing against accidental damage.

12.9.3 Accidental damage: options

(a) Option 1

Option 1 requires the designer to establish that all vertical and horizontal elements are removable one at a time without leading to collapse of any significant portion of the structure. So far as the horizontal members are concerned, this option is superfluous if concrete floor or roof slabs are used, since their structural design must conform to the clause 2.2.2.2(b) of BS 8110: 1985.

(b) Option 3

For the horizontal ties option 3 requirements are very similar to BS 8110: 1985. In addition to this, full vertical ties need to be provided. This option further requires that the minimum thickness of wall should be 150 mm, which makes it a costly exercise. No doubt it would be difficult to provide reinforcements in 102.5 mm wall. However, there could be several ways whereby this problem could be overcome. This option is impracticable in brickwork although possibly feasible for hollow block walls.

(c) Option 2

The only option left is option 2, which can be used in this case. The horizontal ties are required by BS 8110: 1985 to be provided in any case. In addition the designer has to prove that the vertical elements one at a time can be removed without causing collapse.

12.9.4 Design calculations for option 2: BS 5628

(a) Horizontal ties

Basic horizontal tie force, $F_t = 60\,kN$ or $20 + 4N_s$ whichever is less. N_s = number of storeys. Then

$$20 + 4N_s = 20 + 4 \times 7 = 48\,kN < 60\,kN$$

Hence use 48 kN.

(b) Design tie force (table 13, BS 5628)

- Peripheral ties: Tie force, $F_t = 48\,\text{kN}$.

$$\text{As required:} \quad (48 \times 10^3)/250 = 192\,\text{mm}^2$$

  Provide one 16 mm diameter bar as peripheral tie (201 mm^2) at roof and each floor level uninterrupted, located in slab within 1.2 m of the edge of the building.
- Internal ties: Design tie force F_t or $[F_t(G_k + Q_k)/7.5\,] \times L_a/5$ whichever is greater in the direction of span. Tie force

$$F_t = 48\,\text{kN/m} > 48\,\frac{4.8 + 1.5}{7.5} \times \frac{3}{5} = 24.2\,\text{kN/m}$$

  (For the roof the factor G_k is 3.5.) Therefore $F_t = 48\,\text{kN/m}$. (Also note $L_a < 5 \times$ clear height $= 5 \times 2.85 = 14.25\,\text{m}$.) Span of corridor slab is less than 3 m, hence is not considered. Tie force normal to span, $F_t = 48\,\text{kN/m}$.

$$\text{Required} \quad A_s = \frac{(48 \times 1000)}{250} = 192\,\text{mm}^2$$

  Provide 10 mm diameter bar at 400 mm centre to centre in both directions. Area provided 196 mm^2 (satisfactory).

Internal ties should also be provided at each floor level in two directions approximately at right angles. These ties should be uninterrupted and anchored to the peripheral tie at both ends. It will be noted that reinforcement provided for other purposes, such as main and distribution steel, may be regarded as forming a part of, or whole of, peripheral and internal ties (see section 12.10).

(c) Ties to external walls

Consider only loadbearing walls designated as B.

$$\begin{aligned} \text{design tie force} &= 2F_t \text{ or } (h/2.5)\,F_t \quad \text{(whichever is less)} \\ &= 2 \times 48 \text{ or } (2.85/2.5) \times 48\,\text{kN/m} \\ &= 96 \text{ or } 54\,\text{kN/m} \end{aligned}$$

Therefore

design tie force $= 54\,\text{kN/m}$

(d) Tie connection to masonry (Fig. 12.11)

Ignoring the vertical load at the level under consideration, the design characteristic shear stress at the interface of masonry and concrete is

$$\text{shear stress} = \frac{54 \times 10^3 \times \gamma_{mv}}{2 \times 102.5 \times 1000}$$

where

$$\gamma_{mv} = 1.25 \quad \text{(clause 27.4, BS 5628)}$$

$$= \frac{54 \times 10^3 \times 1.25}{2 \times 102.5 \times 1000} = 0.33\,\text{N/mm}^2 < 0.35\,\text{N/mm}^2$$

Hence it is satisfactory, and there is no need to provide external wall ties at any floor level. Further, the vertical load acting at any joint will increase the shear resistance as explained in section 9.5.2(d).

12.9.5 Vertical elements

The designer needs to be satisfied that removal of wall type A, B or C, one at a time, will not precipitate the collapse of the structure beyond specified limits.

To illustrate a method (Sinha and Hendry, 1971), it is assumed that an interior wall type A has been removed from the ground floor of the building. As a result of this incident, the first floor slab will not only deflect due

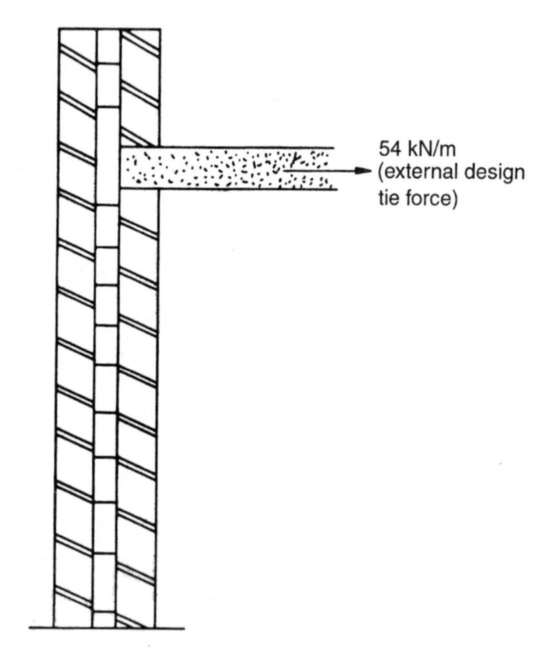

Fig. 12.11 Tie connection.

to the removal of this support but also have to carry the wall load above it without collapsing. As long as every floor takes care of the load imposed on it without collapsing, there is no likelihood of the progressive collapse of the building. This is safer than assuming that the wall above may arch over and transfer the load to the outer cavity and inner corridor walls. Fig. 12.12 shows one of the interior first floor slabs, and the collapse–moment will be calculated by the yield line method. The interior slab has been considered, because this may be more critical than the first interior span, in which reinforcement provided will be higher compared with the interior span. The design calculation for the interior span is given in section 12.10.

The yield-line method gives an upper-bound solution; hence other possible modes were also tried and had to be discarded. It seems that the slab may collapse due to development of yield lines as shown in Fig. 12.12. On removal of wall A below, it is assumed that the slab will behave as simply supported between corridor and outer cavity wall (Fig. 12.1) because of secondary or tie reinforcement.

(a) Floor loading

$$\text{Design dead weight} = \gamma_f \times \text{characteristic dead weight}$$

$$= 1.05 \times 4.8 = 5.04\,\text{kN/m}^2$$

(clause 22d, BS 5628 and Appendix)

$$\text{design imposed load} = \gamma_f \times \text{characteristic imposed load}$$

$$= 1.05 \times 1.5 = 1.58\,\text{kN/m}^2$$

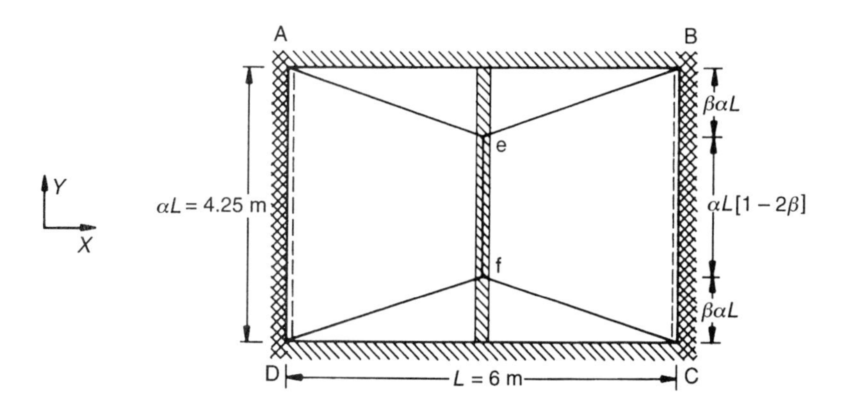

Fig. 12.12 The yield-line patterns at the collapse of the first floor slab under consideration.

Note that γ_f can be reduced to 0.35. According to the code in combination with DL, γ_f factor for LL can be taken as 0.35 in the case of accidental damage. However, it might just be possible that the live load will be acting momentarily after the incident.

$$\text{Total design load } w = 5.04 + 1.58 = 6.62\,\text{kN/m}^2$$

$$\text{Wall load } w' = \gamma_f \times 7.4 \quad \text{(see Table 12.4)}$$

$$= 1.05 \times 7.4 = 7.77\,\text{kN/m}^2$$

(b) Calculation for failure moment

The chosen x and y axes are shown in Fig. 12.12. The yield line ef is given a virtual displacement of unity. External work done $= \Sigma w\delta$, where w is the load and δ is the deflection of the CG of the load. So

$$\Sigma w\delta = 4 \times \tfrac{1}{2} \times w \times \beta\alpha L \times \frac{L}{2} \times \tfrac{1}{3} + 2 \times w \times \frac{L^2}{2}\alpha(1-2\beta) \times \tfrac{1}{2}$$

$$+ 2 \times \tfrac{1}{2} \times w \times \beta\alpha L \times \frac{L}{3} + \frac{2w'\beta\alpha L}{2} + w'\alpha L\,(1-2\beta) \times 1$$

$$= \frac{w\alpha L^2}{6}(3-2\beta) + w'\alpha L - w'\beta\alpha L$$

$$= \frac{6.62 \times 4.25 \times 6}{6}(3-2\beta) + 7.77 \times 4.25 - 7.77 \times 4.25\beta$$

$$= 28.14(3-2\beta) + 33.02 - 33.02\beta$$

$$= 28.14(4.17 - 3.17\beta)$$

Internal work done on the yield lines $= \Sigma(m_x l_x \theta_x + m_y l_y \theta_y)$. So

$$\sum(m_x l_x \theta_x + m_y l_y \theta_y) = \frac{2mL}{\beta\alpha L} + \frac{4m\alpha L}{L/2}$$

$$= \frac{2m}{\beta\alpha} + 8m\alpha$$

$$= \frac{2m}{\beta\alpha}(1 + 4\alpha^2\beta)$$

$$= \frac{2m}{0.71\beta}(1 + 4 \times 0.5\beta) = \frac{2.82m}{\beta}(1+2\beta) \quad (12.58)$$

From equations (12.57) and (12.58)

$$(2.82m/\beta)(1+2\beta) = 28.14(4.17 - 3.17\beta)$$

or

$$m = 9.98\frac{(4.17\beta - 3.17\beta^2)}{1+2\beta} \quad (12.59)$$

For maximum value of moment $dm/d\beta = 0$, from which

$$4.17(1+2\beta) - 8.34\beta - [6.34\beta(1+2\beta) - 3.17\beta^2 \times 2] = 0$$

$$4.17 + 8.34\beta - 8.34\beta - 6.34\beta - 12.68\beta^2 + 6.34\beta^2 = 0$$

$$\beta^2 + \beta = 0.66$$

The positive root of this equation is

$$\beta = 0.45$$

Substituting the value of β in equation (12.59), we get

$$m = \frac{9.98(4.17 \times 0.45 - 3.17 \times 0.45^2)}{1 + 2 \times 0.45} = 6.49 \text{ kNm/m}$$

Then required A_s is

$$A_s = \frac{6.49 \times 10^6}{\gamma_m f_y Z} \quad (\gamma_m = 1, \text{ BS 8110, clause 2.4.4.2})$$

$$= \frac{6.49 \times 10^6}{1 \times 250 \times 90.25} \quad (Z = 90.25, \text{ see Appendix})$$

$$= 287.7 \text{ mm}^2 < 314 \text{ mm}^2 \quad \text{(hence the slab will not collapse)}$$

Owing to removal of support at the ground floor, there will be minimal increase in stresses in the outer cavity and corridor wall. The wall type A(AD and BC in Fig. 12.10) may be relieved of some of the design load, hence no further check is required.

12.10 APPENDIX: A TYPICAL DESIGN CALCULATION FOR INTERIOR-SPAN SOLID SLAB

This is shown in the form of a table (Table 12.5).

Table 12.5 Design calculation for interior span solid slab

BS 8110 ref.	*Calculations*	*Output*
	Durability and fire resistance	
Table 3.4	Nominal cover for mild conditions of exposure = 25 mm	Cover = 25 mm
Table 3.5	Max. fire resistance of 125 mm slab with 25 mm cover = 2 h > 1 h	therefore 1 h fire resistance OK
	Loading	
	Self-weight 0.125 × 24 = 3.0	
	Screed, finish and partition = 1.8	
	= 4.8 kN/m^2	
	Characteristic dead load	$G_k = 4.8$ k/N/m^2
3.5.2.4	Characteristic imposed load (section 12.21)	$Q_k = 1.5$ k/N/m^2
Table 3.13	Design load = (1.4 × 4.8 + 1.6 × 1.5) × 3.0 = 27.36 kN/m width	$F = 27.36$ kN
	Ultimate BM	
Table 3.13	At 1st interior support and mid-span	
	0.063 FL = 0.063 × 27.36 × 3.0 = 5.34 kN m/m	
3.4.4.4	*Reinforcements*	$d = 95$
	1st interior mid-span and support	
	$\frac{M}{f_{cu}bd^2} = \frac{5.34 \times 10^6}{30 \times 10^3 \times 95^2} = 0.02$	
	$z = d\left[0.5 + \left(0.25 - \frac{0.02}{0.9}\right)^{1/2}\right]$	
	$z = 93$ (but $\ngtr 0.95 \times 95 = 90.25$)	
	$A_s = \frac{M}{0.87 f_y z} = \frac{5.34 \times 10^6}{0.87 \times 250 \times 90.25} = 272\,\text{mm}^2$	Top and bottom
Table 3.2.7	Minimum area of steel $= \frac{0.24 \times 1000 \times 125}{100}$	Provide 10Y at 250 mm OK.
	$= 300\,\text{mm}^2 > 272\,\text{mm}^2$	(Area provided 314 mm^2/m)
3.4.6	*Deflection*	
Table 3.10	Basic span/effective depth ratio = 26 maximum	
	$\frac{M}{bd^2} = \frac{5.34 \times 10^6}{10^3 \times (95)^2} = 0.54$	
	$f_s = \frac{5 \times 250 \times 272}{8 \times 314} = 135.4\,\text{N/mm}^2$	
Table 3.11	Modification factor for tension reinforcement = 2.0	
	Therefore allowable span/effective depth ratio = 26 × 2.0 = 52.00	Therefore OK in deflection
	Therefore minimum depth $d = 3000/52.0 = 57.7 < 95$ mm	

Table 12.5 (*Contd*)

BS 8110 ref.	*Calculations*	*Output*
3.12.11.2.7	*Cracking* $3d = 3 \times 95 = 285$ mm Spacing between bars $= 250 - 10 = 240$ mm $< 3d$	Therefore spacing OK
	$h = 1.25 < 250$ mm therefore no further check required	OK
3.27	*Secondary reinforcement* Minimum area of steel $= 0.0024 \times 1000 \times 125$ $= 300\ \text{mm}^2 < 314\ \text{mm}^2$	Bottom and top 10Y at 25 mm
	Spacing between bars $= 250 - 10 = 240 < 250$	OK ($314/\text{mm}^2/\text{m}$)
Table 3.9	*Check for shear* $V = \dfrac{0.255 \times 27.36 \times 10^3}{10^3 \times 95} = 0.16\ \text{N/mm}^2 < v_c$	Therefore OK
	Tie provision Internal ties in both directions (see section 12.8.4) Area required 192 mm^2 (both directions) 70% of primary and secondary reinforcements uninterrupted in both directions will be sufficient $\dfrac{70 \times 314}{100} = 219.8\ \text{mm}^2 > 192\ \text{mm}^2 > 192\ \text{mm}^2$	OK

13

Movements in masonry buildings

13.1 GENERAL

Structural design is primarily concerned with resistance to applied loads but attention has to be given to deformations which result from a variety of effects including temperature change and, in the case of masonry, variations in moisture content. Particular problems can arise when masonry elements are constrained by interconnection with those having different movements, which may result in quite severe stresses being set up. Restraint of movement of a brittle material such as masonry can lead to its fracture and the appearance of a crack. Such cracks may not be of structural significance but are unsightly and may allow water penetration and consequent damage to the fabric of the building. Remedial measures will often be expensive and troublesome so that it is essential for movement to receive attention at the design stage.

13.2 CAUSES OF MOVEMENT IN BUILDINGS

Movement in masonry may arise from the following causes:

- Moisture changes
- Temperature changes
- Strains due to applied loads
- Foundation movements
- Chemical reactions in materials

13.2.1 Moisture movements

Dimensional changes take place in masonry materials with change in moisture content. These may be irreversible following manufacture – thus clay bricks show an expansion after manufacture whilst concrete and calcium silicate products are characterized by shrinkage. All types of masonry exhibit reversible expansion or shrinkage with change in moisture content at all stages of their existence. Typical values are shown in Table 13.1.

13.2.2 Thermal movements

Thermal movements depend on the coefficient of expansion of the material and the range of temperature experienced by the building element. Values of the coefficient of expansion are indicated in Table 13.1 but estimation of the temperature range is complicated depending as it does on other thermal properties such as absorptivity and capacity and incident solar radiation. The temperature range experienced in a heavy exterior wall in the UK has been given as −20°C to +65°C but there are likely to be wide variations according to colour, orientation and other factors.

13.2.3 Strains resulting from applied loads

Elastic and creep movements resulting from load application may be a factor in high-rise buildings if there is a possibility of differential movement between a concrete or steel frame and masonry cladding or infill. Relevant values of elastic modulus and creep coefficients are quoted in Chapter 4.

13.2.4 Foundation movements

Foundation movements are a common cause of cracking in masonry walls and are most often experienced in buildings constructed on clay soils which are affected by volume changes consequent on fluctuation in soil moisture content. Soil settlement on infilled sites and as a result of mining operations is also a cause of damage to masonry walls in certain areas. Where such problems are foreseen at the design stage suitable

Table 13.1 Moisture and thermal movement indices for masonry materials, concrete and steel

Material	*Reversible moisture movement* (%)	*Irreversible moisture movement* (%)	*Coefficient of thermal expansion* (10^{-6}/°C)
Clay brickwork	0.02	+0.02–0.07	4–8
Calcium silicate brickwork	0.01–0.05	−0.01–0.04	8–14
Concrete brick- or blockwork	0.02–0.04	−0.02–0.06	7–14
Aerated, autoclaved blockwork	0.02–0.03	−0.05–0.09	8
Dense aggregate concrete	0.02–0.10	−0.03–0.08	10–14
Steel	–	–	12

precautions can be taken in relation to the design of the foundations, the most elementary of which is to ensure that the foundation level is at least 1m below the ground surface. More elaborate measures are of course required to cope with weak soils or mining subsidence.

13.2.5 Chemical reactions in materials

Masonry materials are generally very stable and chemical attack in service is exceptional. However, trouble can be experienced as the result of sulphate attack on mortar and on concrete blocks and from the corrosion of wall ties or other steel components embedded in the masonry.

Sulphate solution attacks a constituent of cement in mortar or concrete resulting in its expansion and disintegration of the masonry. The soluble salts may originate in ground water or in clay bricks but attack will only occur if the masonry is continuously wet. The necessary precaution lies in the selection of masonry materials, or if ground water is the problem, in the use of a sulphate-resistant cement below damp-proof course level.

13.3 HORIZONTAL MOVEMENTS IN MASONRY WALLS

Masonry in a building will rarely be free to expand or contract without restraint but, as a first step towards appreciating the magnitude of movements resulting from moisture and thermal effects, it is possible to deduce from the values given in Table 13.1 the theoretical maximum change in length of a wall under assumed thermal and moisture variations. Thus the maximum moisture movement in clay brick masonry could be an expansion of 1mm in 1m. The thermal expansion under a temperature rise of 45°C could be 0.3 mm so that the maximum combined expansion would be 1.3 mm per metre. Aerated concrete blockwork on the other hand shrinks by up to 1.2 mm per metre and has about the same coefficient of thermal expansion as clay masonry so that maximum movement would be associated with a fall in temperature.

Walls are not, in practical situations, free to expand or contract without restraint but these figures serve to indicate that the potential movements are quite large. If movement is suppressed, very large forces can be set up, sufficient to cause cracking or even more serious damage. Provision for horizontal movement is made by the selection of suitable materials, the subdivision of long lengths of wall by vertical movement joints and by the avoidance of details which restrain movement and give rise to cracking.

The spacing of vertical movement joints is decided on the basis of empirical rules rather than by calculation. Such joints are filled with a compressible sealant and their spacing will depend on the masonry material. An upper limit of 15 m is appropriate in clay brickwork, 9 m in calcium silicate brickwork and 6 m in concrete blockwork. Their width in millimetres should be about 30% more than their spacing in metres. Location in the building will depend on features of the building such as intersecting walls and openings. It should be noted that the type of mortar used has an important influence on the ability of masonry to accommodate movement: thus a stone masonry wall in weak lime mortar can be of very great length without showing signs of cracking. Brickwork built in strong cement mortar, on the other hand, will have a very much lower tolerance of movement and the provision of movement joints will be essential.

Certain details, such as short returns (Fig. 13.1) are particularly vulnerable to damage by moisture and thermal expansion. Similar damage can result from shrinkage in calcium silicate brickwork or concrete blockwork. Parapet walls are exposed to potentially extreme variations of temperature and moisture and their design for movement therefore requires special care. A considerable amount of guidance on these points is provided in BS 5628: Part 3.

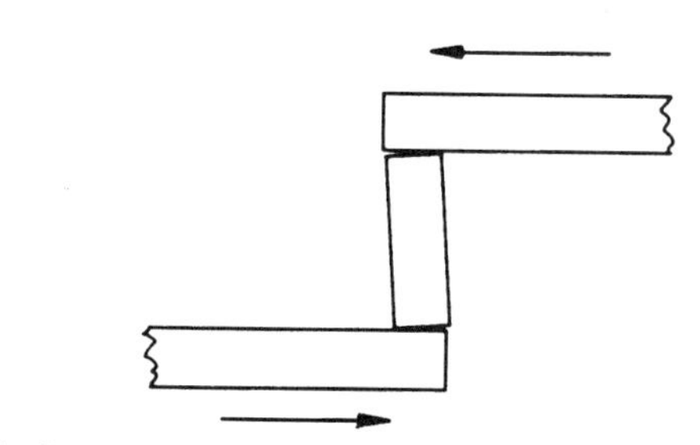

(a) Cracking at a short return due to masonry expansion

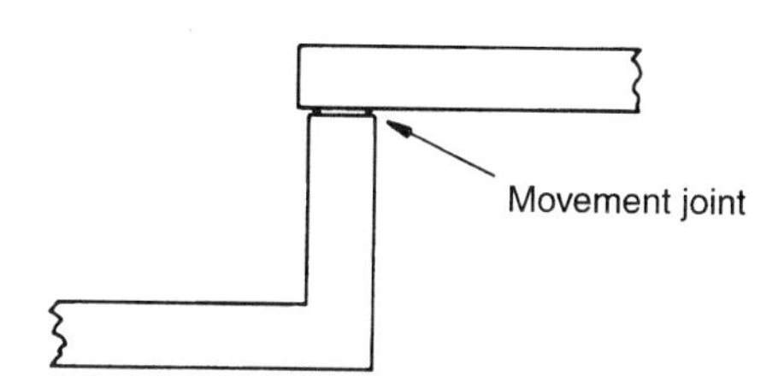

(b) Cracking avoided by insertion of a movement joint

Fig. 13.1 Cracking at a short return in brick masonry.

13.4 VERTICAL MOVEMENTS IN MASONRY WALLS

Vertical movements in masonry are of the same order as horizontal movements but stress-related movements in multi-storey walls will be of greater significance. Vertical movements are of primary importance in the design of cavity walls and masonry cladding to reinforced concrete or steel-framed buildings. This is because the outer leaf of masonry will generally have different characteristics to those of the inner leaf or structure and will be subjected to different environmental conditions. This will result in differential movements between the outer leaf and the inner wall which could lead to loosening of wall ties or fixtures between them or in certain circumstances to serious damage to the masonry cladding.

To avoid problems from this cause, BS 5628: Part 1 states that the outer leaf of an external cavity wall should be supported at intervals of not more than three storeys or 9 m (12 m in a four-storey building). Alternatively, the relative movement between the inner wall and the outer leaf may be calculated and suitable ties and details provided to allow such movement to take place.

The approximate calculation of vertical movements in a multi-storey, non-loadbearing masonry wall may be illustrated by the following example, using hypothetical values of masonry properties. Height of wall = 24 m. Number of storeys = 8.

- *Moisture movements.* Irreversible shrinkage of masonry, 0.005 25%. Shrinkage in height of wall, $0.0000525 \times 24 \times 10 = 1.26$ mm. Reversible moisture movement from dry to saturated state, $\pm 0.04\%$. Moisture movement taking place depends on moisture content at time of construction. Assuming 50% saturation at this stage reversible movement may be

$$0.5 \times 0.0004 \times 24 \times 10^3 = \pm 4.8 \text{ mm.}$$

Table 13.2 Elastic and creep deformations

Storey	*Average stress* (kN/m²)	*Strain* ($\times 10^{-6}$)	*Compression in storey* (mm)	*Cumulative compression* (mm)	*Creep compression* (mm)
8	56.1	26.7	0.08	5.12	7.68
7	168.4	80.2	0.24	5.04	7.56
6	280.7	133.7	0.40	4.80	7.20
5	393.0	187.1	0.56	4.40	6.60
4	505.3	240.6	0.72	3.84	5.76
3	617.5	294.0	0.88	3.12	4.68
2	729.8	347.5	1.04	2.24	3.36
1	842.2	401.0	1.20	1.20	1.8

- *Elastic and creep movements.* Elastic modulus of masonry, 2100 N/mm^2. Creep deformation, 1.5 × elastic deformation. Elastic and creep deformations, due to self-weight, at each storey level are tabulated in Table 13.2.
- *Thermal movement.* Coefficient of thermal expansion, 10×10^{-6} per °C. Assumed temperature at construction, 10°C. Minimum mean temperature of wall, −20°C. Maximum mean temperature of wall, 50°C. Range in service from 10°C, −10°C to +40°C. Overall contraction of wall

$$30 \times 10 \times 10^{-6} \times 24 \times 10^3 = 7.2 \text{ mm}$$

Overall expansion of wall

$$40 \times 10 \times 10^{-6} \times 24 \times 10^3 = 12.8 \text{ mm}$$

The maximum movement at the top of the wall due to the sum of these effects is as follows:

	Outer wall	*Inner wall*
Irreversible moisture movement	−1.3	+9.6
Reversible moisture movement	−4.8	–
Elastic deformation	−5.0	−0.8
Creep	−7.7	−1.2
Thermal movement	−7.2	+4.8
	−26.0 mm	+12.4 mm

Shown in the right-hand column are comparable figures for a clay brickwork inner wall which would show irreversible moisture expansion rather than contraction and would reach a stable moisture state after construction so that irreversible moisture movement has been omitted in this case. The wall would also experience a rise in temperature when the building was brought into service and thus thermal expansion would take place. In this example there would be a possible differential movement at the top of the wall of 38.7 mm but as movements are cumulative over the height of the wall it is of interest to calculate the relative movements at storey levels.

This calculation is set out in detail for the outer wall in Table 13.3. The corresponding figures for the inner wall and the relative movements which would have to be accommodated at each storey level are also shown in the table and graphically in Fig. 13.2. Note that if the walls are built at the same time the differential movement due to elastic compression is reduced since the compression below each level will have taken place before the ties are placed. Thus the relative wall tie movement due to elastic compression at the top level will be zero.

Table 13.3 Masonry outer wall – clay brickwork inner wall: relative wall tie movements at storey levels

Storey	8	7	6	5	4	3	2	1
Shrinkage	−1.3	−1.1	−1.0	−0.8	−0.6	−0.5	−0.3	0
Rev. moisture movement	−4.8	−4.2	−4.0	−3.0	−2.4	−1.8	−1.2	−0.6
Elastic compression	0	−0.3	−0.4	−0.6	−0.8	−0.9	−1.1	−1.2
Creep	−7.4	−6.7	−5.9	−5.2	−4.5	−3.8	−3.0	−2.3
Thermal	−7.2	−6.2	−5.3	−4.3	−3.4	−2.4	−1.5	−0.5
Total	−20.7	−18.6	−16.6	−14.9	−11.7	−9.4	−7.1	−4.6
Total movement in brickwork inner wall	+14.4	+12.5	+10.5	+8.6	+6.7	+4.8	+2.8	+0.9
Movement across wall ties	35.1	31.1	27.1	23.3	18.4	14.2	9.9	5.3

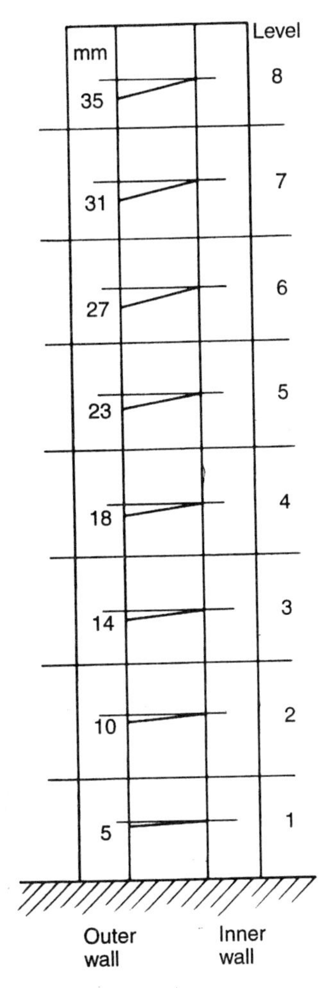

Fig. 13.2 Relative wall tie movements.

Movements across the cavity of the order shown would require the use of special wall ties, many varieties of which are commercially available. It is also necessary to allow for differential movements across the cavity at window openings and at the roof level requiring careful detailing to preserve water exclusion as well as permitting free movement.

As suggested above, differential movement between the leaves of a cavity wall or between masonry cladding and the main structure of a building will depend on the characteristics of both. If the main structure is a steel frame the only significant movement in it will be the result of temperature change from that assumed at construction to a maximum in

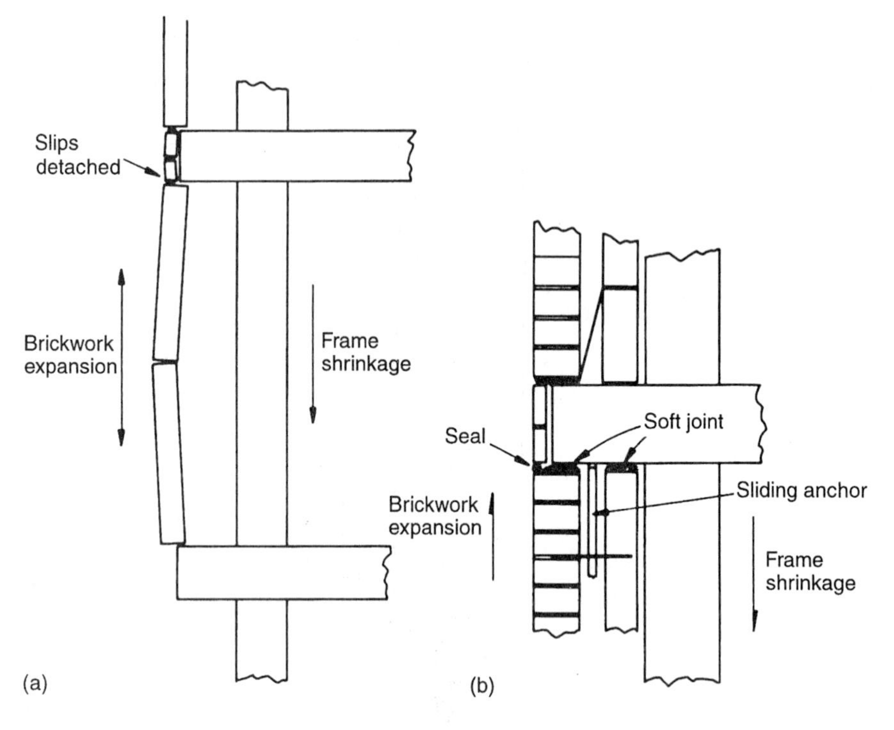

Fig. 13.3 (a) Bowing of infill wall and detachment of brick slips as a result of frame shrinkage. (b) Detail of horizontal movement joint to avoid damage of the kind shown in (a).

service. A concrete main structure will, however, develop shrinkage and creep strains after completion which will have to be allowed for in estimating differential movements relative to a masonry cladding. If masonry cladding is built between concrete floor slabs, as in Fig. 13.3(a), a serious problem can be created if the masonry expands and the concrete frame shrinks unless this relative movement is allowed for by suitable detailing as in Fig. 13.3(b).

Notation

BS 5628

A	cross-sectional area of masonry (mm^2)
A_{ps}	cross-sectional area of prestressing steel (mm^2)
A_s	cross-sectional area of primary reinforcing steel (mm^2)
A_{sv}	cross-sectional area of reinforcing steel resisting shear forces (mm^2)
A_{s1}	area of compression reinforcement in the most compressed face (mm^2)
A_{s2}	area of reinforcement in the least compressed face (mm^2)
a	shear span (mm^2)
a_v	distance from face of support to the nearest edge of a principal load (mm)
b	width of section (mm)
b_c	width of compression face midway between restraints (mm)
b_1	width of section at level of the tension reinforcement (mm)
c	lever arm factor
d	effective depth (mm)
d_c	depth of masonry in compression (mm)
d_1	depth from the surface to the reinforcement in the more highly compressed face (mm)
d_2	depth of the centroid of the reinforcement from the least compressed face (mm)
E_c	modulus of elasticity of concrete (kN/mm^2)
E_m	modulus of elasticity of masonry (kN/mm^2)
E_m, E_b	modulus of elasticity of mortar and brick (kN/mm^2)
E_s	modulus of elasticity of steel (kN/mm^2)
E_x, E_y	modulus of elasticity in x and y direction (kN/mm^2)
e	eccentricity
e_a	additional eccentricity due to deflection in walls
e_m	the larger of e_x or e_t
e_t	total design eccentricity in the mid-height region of a wall
e_x	eccentricity at top of a wall

F_k	characteristic load
F_t	tie force
f_b	characteristic anchorage bond strength between mortar or concrete infill and steel (N/mm^2)
f_{ci}	strength of concrete at transfer (N/mm^2)
f_k	characteristic compressive strength of masonry (N/mm^2)
f_{kx}	characteristic flexural strength (tension) of masonry (N/mm^2)
f_m	masonry strength
f_{pb}	stress in tendon at the design moment of resistance of the section (N/mm^2)
f_{pe}	effective prestress in tendon after all losses have occurred (N/mm^2)
f_{pu}	characteristic tensile strength of prestressing tendons (N/mm^2)
f_s	stress in the reinforcement (N/mm^2)
f_{su}	stress in steel at failure
f_{s1}	stress in the reinforcement in the most compressed face (N/mm^2)
f_{s2}	stress in the reinforcement in the least compressed face (N/mm^2)
f_v	characteristic shear strength of masonry (N/mm^2)
f_y	characteristic tensile strength of reinforcing steel (N/mm^2)
G_k	characteristic dead load
g_A	design vertical load per unit area
g_d	design vertical dead load per unit area
h	clear height of wall or column between lateral supports
h_a	clear height of wall between concrete surfaces or other construction capable of providing adequate resistance to rotation across the full thickness of a wall
h_{ef}	effective height or length of wall or column
h_L	clear height of wall to point of application of a lateral load
K	stiffness coefficient
k	multiplication factor for lateral strength of axially loaded walls
L	length
L_a	span in accidental damage calculation
M	bending moment due to design load (N mm)
M_a	increase in moment due to slenderness (N mm)
M_d	design moment of resistance (N mm)
M_x	design moment about the x axis (N mm)
$M'_{x'}$	effective uniaxial design moment about the x axis (N mm)
M_y	design moment about the y axis (N mm)
M'_y	effective uniaxial design moment about the y axis (N mm)
N	design axial load (N)

N_d	design axial load resistance (N)
N_{dz}	design axial load resistance of column, ignoring all bending (N)
P, P_e	prestressing forces
p	overall section dimension in a direction perpendicular to the x axis (mm)
Q	moment of resistance factor (N/mm^2)
Q_k	characteristic imposed load (N)
q	overall section dimension in a direction perpendicular to the y axis (mm)
q_{lat}	design lateral strength per unit area
q_0, q_1, q_2	transverse or lateral pressure
t	overall thickness of a wall or column (mm)
t_{ef}	effective thickness of a wall or column (mm)
t_f	thickness of a flange in a pocket-type wall (mm)
V	shear force due to design loads (N)
v, v_h	shear stress due to design loads (N/mm^2)
W_k	characteristic wind load (N)
Z, Z_1, Z_2	section modulus (mm^3)
z	lever arm (mm)
α	bending moment coefficient for laterally loaded panels in BS 5628
β	capacity reduction factor for walls allowing for effects of slenderness and eccentricity
γ_f	partial safety factor for load
γ_m	partial safety factor for material
γ_{mb}	partial safety factor for bond strength between mortar or concrete infill and steel
γ_{mm}	partial safety factor for compressive strength of masonry
γ_{ms}	partial safety factor for strength of steel
γ_{mv}	partial safety factor for shear strength of masonry
ε	strain as defined in text
λ_1, λ_2	stress block factors
μ_f	coefficients of friction
ν_b, ν_m	Poisson's ratio for brick and mortar
ν_x, ν_y	Poisson's ratios in x and y direction
μ	orthogonal ratio
ρ	A_s/bd
σ	compressive stress
σ_b	compressive stress in brick
σ_m	compressive stress in mortar or in masonry
σ_s	stress in steel
ϕ	creep loss factor

EC6 (WHERE DIFFERENT FROM BS 5628)

e_a	eccentricity resulting from construction inaccuracies
e_{hi}	eccentricity resulting from lateral loads
e_i	eccentricity at top or bottom of wall
e_k	eccentricity allowance for creep
e_{mk}	eccentricity at mid-height of wall
f_b	normalized unit compressive strength
f_m	specified compressive strength of mortar
f_{tk}	characteristic tensile strength of steel
f_{vk}	characteristic shear strength of masonry
f_{yk0}	shear strength of masonry under zero compressive stress
f_{yk}	characteristic yield strength of steel
I	second moment of area
K	constant concerned with characteristic strength of masonry
k	stiffness factor
L	distance between centres of stiffening walls
l_c	compressed length of wall
l_e	effective length or span
M_i	design bending moment at top or bottom of a wall
M_m	design bending moment at mid-height of a wall
M_{RD}	design bending moment of a beam
N_i	design vertical load at top or bottom of a wall
N_{RD}	design vertical load resistance per unit length
W	distributed load on a floor slab
γ_G	partial safety factor for permanent actions
γ_Q	partial safety factor for variable actions
γ_P	partial safety factor for prestressing
γ_s	partial safety factor for steel
δ	shape factor for masonry units
$\Phi_{i,m}$	capacity reduction factor allowing for the effects of slenderness and eccentricity
Φ_∞	final creep coefficient
ρ_n	reduction factor for wall supported on vertical edges
σ_d	design compressive stress normal to the shear stress

Definition of terms used in masonry

bed joint horizontal mortar joint
bond (1) pattern to which units are laid in a wall, usually to ensure that cross joints in adjoining courses are not in vertical alignment; (2) adhesion of bricks and mortar
cavity wall two single-leaf walls spaced apart and tied together with wall ties
chase a groove formed or cut in a wall to accommodate pipes or cables
collar joint vertical joint in a bonded wall parallel to the face
column an isolated vertical compression member whose width is not less than four times its thickness
course a layer of brickwork including a mortar bed
cross joint a vertical joint at right angles to the face of a wall
efflorescence a deposit of salts on the surface of a wall left by evaporation
fair-faced a wall surface carefully finished with uniform jointing and even setting of bricks for good appearance
frog an indentation on the bedding surface of a brick
grout a mix consisting of cement, lime, sand and pea gravel with a sufficiently large water content to permit its being poured or pumped into cavities or pockets without the need for subsequent tamping or vibration
header a unit laid with its length at right angles to the face of the wall
leaf a wall, forming one skin or cavity
movement joint a joint designed to permit relative longitudinal movement between contiguous sections of a wall in a building
panel an area of brickwork with defined boundaries, usually applied to walls resisting predominantly lateral loads
perpend the vertical joint in the face of a wall
pier a compression member formed by a thickened section of a wall
pointing the finishing of joints in the face of a wall carried out by raking out some of the mortar and re-filling either flush with the face or recessed in a particular way

racking shear a horizontal, in-plane force applied to a wall
shear wall a wall designed to resist horizontal, in-plane forces, e.g. wind loads
spalling a particular mode of failure of brickwork in which chips or large fragments generally parallel to the face of the brick are broken off
stretcher a unit laid with its length parallel to the face of the wall

References and further reading

Coull, A. and Stafford-Smith, B. S. (eds) (1967) *Tall Buildings – Proc. Symp. on Tall Buildings*, Pergamon, Oxford

Curtin, W. G., Shaw, G., Beck, J. G. and Bray, W. A. (1995) *Structural Masonry Designer's Manual*, Blackwell, Oxford

Davies, S. R. (1995) *Spreadsheets in Structural Design*, Longman, Harlow

Hendry, A. W. (1990) *Structural Masonry*, Macmillan, Basingstoke

Hendry, A. W. (ed.) (1991) *Reinforced and Prestressed Masonry*, Longman, Harlow

Hendry, A. W., Sinha, B. P. and Maurenbrecher, A. H. P. (1971) Full-scale tests on the lateral strength of brick cavity-walls without precompression. *Proc. 4th Symp. on Load-bearing Brickwork*, British Ceramic Society, Stoke-on-Trent, pp. 141–64

Khalaf, F. M. (1991) *Ph.D. Thesis*, University of Edinburgh

Kukulski, W. and Lugez, J. (1966) *Résistance des Murs en Béton non Armée Soumis à des Charges Verticales*, Cahiers CSTB, No. 681

Page, A. W. and Hendry, A. W. (1987) Design rules for concentrated loads on masonry, *Structural Engineer*, **66**, 273–81

Pedreschi, R. F. and Sinha, B. P. (1985) Deformation and cracking of post-tensioned brick-work beams, *Structural Engineer*, **63B** (4), December, 93–100

Riddington, J. R. and Stafford-Smith, B. S. (1977) Analysis of infilled frames subject to racking – with design recommendations, *Structural Engineer*, **55**(6) June, 263–8

Roberts, J. J., Tovey, A. K., Cranston, W. B. and Beeby, A. W. (1983) *Concrete Masonry Designer's Handbook*, Viewpoint, Leatherhead

Sinha, B. P. (1978) A simplified ultimate load analysis of laterally loaded model orthotropic brickwork panels of low tensile strength, *Structural Engineer*, **50B**(4), 81–4

Sinha, B. P. (1980) An ultimate load analysis of laterally loaded brickwork panels, *Int. J. Masonry Construction*, **1**(2), 5741

Sinha, B. P. and Hendry, A. W. (1971) The stability of a five-storey brickwork cross-wall structure following removal of a section of a main load-bearing wall, *Structural Engineer*, **49**, October, 467–74

Stafford-Smith, B. S. and Riddington, J. R. (1977) The composite behaviour of elastic wall–beam systems, *Proc. Inst. Civ. Eng. (Part 2)*, **63**, June, 377–91

Wood, R. H. (1952) *Studies in Composite Construction*, Part 1, *The Composite Action of Brick Panel Walls Supported on Reinforced Concrete Beams*, National Building Studies Research, Paper 13

Wood, R. H. (1978) Plasticity, composite action and collapse design of unreinforced shear wall panels in frames, *Proc. Inst. Civ. Eng. (Part 2)*, **65**, June, 381–441

Wood, R. H. and Simms, L. G. (1969) *A Tentative Design Method for the Composite Action of Heavily Loaded Brick Panel Walls Supported on Reinforced Concrete Beams*, Building Research Station, CP26/29

Index